THE MILLETS
AND MINOR CEREALS

A Bibliography of the World Literature
on Millets pre-1930 and 1964-69; and of
All Literature on Other Minor Cereals

by

KENNETH O. RACHIE
The Rockefeller Foundation

The Scarecrow Press, Inc.
Metuchen, N. J.   1974

The present work supplements The Millets: A Bibliography of the World Literature covering the years 1930-1963, published by the Scarecrow Press in 1967 (L. C. Card no. 66-30356), and covering seven millet species only.

Library of Congress Cataloging in Publication Data

Rachie, Kenneth O
    The millets and minor cereals; a bibliography of the world literature on millets, pre-1930 and 1964-1969, and of all literature on other minor cereals.

    "The present work supplements The millets: a bibliography of the world literature covering the years 1930-1963, published ... in 1967."
    1. Millet--Bibliography. I. George Washington University, Washington, D. C. Biological Sciences Communication Project. The millets. II. Title.
Z5074.M5R32   016.633'171   73-22142
ISBN 0-8108-0700-9

TABLE OF CONTENTS

iii

PREFACE

The improvement of millets and certain minor cereals as
potential food sources has been largely neglected by comparison
with rice, wheat and maize, the preferred grains of advanced regions. Nevertheless, there is increasing recognition that the
"elite" cereals do not necessarily perform well in all environments, meet all requirements, are the most nutritious, nor produce the greatest amount of food on a unit of land/time/nutrients/
water basis. Moreover, the expanding cultivation of crops on mar-
ginal lands places emphasis on hardier species and those better
adapted to conditions of stress like heat, drought, impoverished
soils, alkalinity, salinity, or acidity.

Recognizing the future potential of grain crops like sorghum
and millets in developing countries--particularly in tropical and
sub-tropical regions--the Rockefeller Foundation is assembling the
world's literature in the form of bibliographies and a comprehen-
sive series of monographs. These are intended to serve the needs
of scientists, students, crop specialists and others concerned with
the improvement of the millets and minor cereals. The first bibliography, The Millets, was organized and written by the Biological
Sciences Communication Project of the George Washington University
and published by the Scarecrow Press of Metuchen, New Jersey,
in 1967. It contained 1225 bibliographies on seven species covering
the years 1930 to 1963.

The present bibliography is intended as a sequel to that work
in that it covers the period prior to 1930 and from 1964 to mid-
1969 for the same seven species; and all literature for seven addi-
tional species up to mid-1969. The 1439 references in this compilation directly concern the following crops (*included in the 1967 bibliography):

COMMON NAME	SCIENTIFIC
*1. Bajra, pearl or bulrush millet	Pennisetum typhoides Stapf and Hubb.
*2. Italian or foxtail millet	Setaria italica Beauv.
*3. Proso or common millet	Panicum miliaceum Linn.
*4. Little millet	Panicum miliare Lam.
*5. Ragi or finger millet	Eleusine coracana Gaertn.
*6. Koda or ditch millet	Paspalum scrobiculatum Linn.
*7. Japanese barnyard millet	Echinochloa frumentacea (Roxb.) Link.
8. Jungle rice or shama millet	Echinochloa colona (L.) Link.
9. Australian millet	Echinochloa decompositum
10. Browntop millet	Brachiaria ramosa (Linn.) Stapf.
11. Teff	Eragrostis tef (Zucc.) Trotter
12. Fonio or hungry rice	Digitaria iburua Stapf.
13. Fonio or hungry rice	Digitaria exilis Stapf.
14. Adlay or Job's tears	Coix lachryma-jobi Linn.

This compilation was initiated at the suggestion of Dr. Sterling Wortman, vice president of The Rockefeller Foundation, New York. Much encouragement and support have also been received from Drs. J. J. McKelvey and D. P. Parker, both associate directors for agricultural sciences, The Rockefeller Foundation. Deep appreciation is also extended to Mrs. Joan Holliday of Kampala, Uganda, who assisted with organizing, indexing and checking the manuscript.

Kenneth O. Rachie

SERIAL TITLES ABBREVIATIONS

Acta Agr. Sin.
  Acta Agriculturae Sinica
Acta Agrar. Silv., Krakow:
  Ser. Agrar.
  Acta Agraria et Silvestria.
  Krakow. Series Agraria
Acta Agrar. Silv., Krakow:
  Ser. Roln.
  Acta Agraria et Silvestria.
  Krakow. Series Rolnicza
Acta Bot. Sin.
  Acta Botanica Sinica
Acta Pedol. Sin.
  Acta Pedologica Sinica
Acta Phytopath. Sin.
  Acta Phytopathologica Sinica
Acta Phytophylac. Sin.
  Acta Phytophylacica Sinica
Advancing Frontiers. Plant Sci.
  Advancing Frontiers of Plant
  Sciences
Agr. Bul. Fed. Malay States
  Agricultural Bulletin. Fed-
  erated Malay States
Agr. Meteorol.
  Agricultural Meteorology
Agra Univ. J. Res. Sci.
  Agra University Journal of
  Research, Science
Agric. Anim. Husb. India
  Agriculture and Animal Hus-
  bandry in India
Agric. Gaz. N. S. W.
  Agricultural Gazette of New
  South Wales
Agric. J. India
  Agricultural Journal of India
Agric. Livestock India
  Agriculture and Live-Stock in
  India
Agric. Res. (India)
  Agricultural Research (India)

Agric. Res. (Pretoria)
  Agricultural Research (Pretoria)
Agric. Res. (Wash.)
  Agricultural Research (Wash-
  ington)
Agric. Trop.
  Agricultura Tropical
Agriculture (London)
  Agriculture (London)
Agrobiologiya
  Agrobiologiya
Agrokhimiya
  Agrokhimiya
Agron. Abstr.
  Agronomy Abstracts
Agron. J.
  Agronomy Journal
Agron. Trop.
  L'Agronomie Tropicale
Allahabad Fmr.
  Allahabad Farmer
Amer. Chem. J.
  American Chemical Journal
Amer. J. Bot.
  American Journal of Botany
An. Congr. Soc. Bot. Brazil
  Anais do Congresso Nacional.
  Sociedade Botanica do
  Brasil
An. Fac. Med. Univ. Sao Paulo
  Anais da Faculdade de Medi-
  cina da Universidade de Sao
  Paulo
Analyst
  Analyst
Andhra Agric. J.
  Andhra Agricultural Journal
Angew. Bot.
  Angewandte Botanik
Ann. Appl. Biol.
  Annals of Applied Biology
Ann. Arid Zone
  Annals of Arid Zone

Ann. Bot.
  Annals of Botany
Ann. Centre Rech. Agron.
  Bambey
  Annales du Centre de Re-
    cherches Agronomiques de
    Bambey au Senegal
Ann. Mo. Bot. Gdn.
  Annals of Missouri Botanical
    Garden
Ann. Nutrition Alimentation
  Annales de la Nutrition et
    de l'Alimentation
Ann. Phytopath. Soc. Japan
  Annals of the Phytopatholog-
    ical Society of Japan
Ann. Rep. Fla. Agric. Exp. Sta.
  Annual Report. Florida Agri-
    cultural Experiment Station
Ann. Rep. Miss. Agric. Exp.
    Sta.
  Annual Report. Missouri
    Agricultural Experiment
    Station
Ann. Rep. Ont. Agr. Coll.
    Exp. Farm
  Annual Report of the Ontario
    Agricultural College and
    Experimental Farm
Ann. Sper. Agr.
  Annali della Sperimentazione
    Agraria
Ann. Univ. Sofia. Fac. Agric.
  Annuaire de l'Universite de
    Sofia. Faculte d'Agriculture
Annls. Univ. Mariae Curie-
    Sklodowska Sect. E Agric.
  Annales Universitatis Mariae
    Curie-Sklodowska. Sect.
    E Agriculture
Annu. Rep. Dep. Agric. Orange
    River Colony
  Annual Report of the Depart-
    ment of Agriculture,
    Orange River Colony
Arch. Pharm. u. Ber. Deutsch.
    Pharmazeut. Gesell.
  Archiv der Pharmazie und
    Berichte der Deutschen
    Pharmazeutischen Gesell-
    schaft
Arch. Venezol. Nutricion

Archives Venezolanos de Nutri-
    cion
Arkansas Agr. Exp. Stn. Rep.
    Ser.
  Arkansas. Agricultural Experi-
    ment Station. Report Series
Arkansas Sta. Bull.
  Arkansas Agricultural Experi-
    ment Station Bulletin
Aryaswapatra
  Arya Swapatra
Atti Soc. Pelorit. Sci. Fis. Mat.
    Nat.
  Atti. Societa Peloritana di
    Scienze Fisiche, Matema-
    tiche e Naturali
Aust. J. Biol. Sci.
  Australian Journal of Bio-
    logical Sciences
Aust. J. Bot.
  Australian Journal of Botany
Aust. J. Exp. Agr. Anim. Husb.
  Australian Journal of Experi-
    mental Agriculture and
    Animal Husbandry
Australian Reference Newsletter
  Australian Reference Newsletter

Balwant Vidyapeeth J. Agric. Sci.
    Res.
  Balwant Vidyapeeth Journal of
    Agricultural & Scientific
    Research
Beitr. Trop. Subtrop. Landwirt.
    Tropenveterinarmed.
  Beiträge zur Tropischen und
    Subtropischen Landwirtschaft
    und Tropenveterinärmedizin
Ber. Dt. Bot. Ges.
  Berichte der Deutschen
    Botanischen Gesellschaft
Ber. Ohara Inst.
  Berichte. Ohara Institut für
    Landwirtschaftliche Biologie,
    Okayama Universität
Bibljot. Pulaw.
  Bibljoteka Pulawska
Biul. Inst. Hodowl. Aklimatyz.
    Roslin
  Biuletyn Instytutu Hodowla
    Aklimatyzacja Roslin
Bodenkultur
  Bodenkultur

Bol. Industr. Animal
  Boletim de Industria Animal
Bol. Lab. Ent. Agr. Portici
  Bollettino del Laboratorio di
  Entomologia Agraria
  'Filippo Silvestri.' Portici
Boll. Soc. Bot. Ital.
  Bollettino della Societa
  Botanica Italiana
Bot. Bull. Acad. Sin.
  Botanical Bulletin of
  Academia Sinica
Bot. Gaz.
  Botanical Gazette
Bothalia
  Bothalia
Bragantia
  Bragantia
Brittonia
  Brittonia
Bull. Agric. Congo
  Bulletin Agricole du Congo
Bull. Agric. Congo Belge
  Bulletin Agricole du Congo
  Belge
Bull. Agric. Res. Inst. Pusa.
  Bulletin. Agricultural Re-
  search Institute, Pusa
Bull. Agron. Minist. Fr.
  d'Out. Mer
  Bulletin Agronomique. Min-
  istere de la France
  d'Outre Mer
Bull. Anciens Eleves Ecole
  Francaise Meunerie
  Bulletin des Anciens Eleves
  de l'Ecole Francaise de
  Meunerie
Bull. Appl. Bot. Genet-Pl.
  Breed.
  Trudy po Prikladnoi Botanike,
  Genetike i Selektsii
Bull. Bot. Surv. India.
  Bulletin of the Botanical
  Survey of India
Bull. Dep. Agric. Bombay
  Bulletin. Department of Agri-
  culture, Bombay (State)
Bull. Dep. Agric. Mysore
  Entomol. Ser.
  Bulletin. Department of Agri-

culture, Mysore State.
  Entomology Series
Bull. Ent. Res.
  Bulletin of Entomological Re-
  search
Bull. Fac. Agric. Yamaguchi
  Bulletin of the Faculty of Agri-
  culture, Yamaguchi Univer-
  sity
Bull. Grain Technol.
  Bulletin of Grain Technology
Bull. Indian Phytopath. Soc.
  Bulletin. Indian Phytopatholog-
  ical Society
Bull. Ineac
  Bulletin. Institut National pour
  l'Etude Agronomique du
  Congo
Bull. Intl. Acad. Sci. Cracovie,
  Cl. Math. et Nat.
  Bulletin International de
  l'Academie des Sciences et
  des Lettres de Cracovie.
  Classe des Sciences Mathe-
  matique et Naturelles
Bull. Natl. Hyg. Lab. (Tokyo)
  Bulletin of the National Hy-
  gienic Laboratory. Tokyo
Bull. Polytech. Inst. Teflis
  Bulletin of the Polytechnical
  Institute of Tiflis
Bull. Shikoku Agric. Exp. Stn.
  Bulletin of the Shikoku Agri-
  cultural Experiment Station
Bull. Soc. Bot. Fr.
  Bulletin. Societe Botanique de
  France
Bull. Soc. His. Natur. Toulouse
  Bulletin de la Societe d'Histoire
  Naturelle de Toulouse
Bull. Stat. Agron. Reunion
  Travaux Techniques
  Bulletin dela Station Agrono-
  mique, Réunion (Travaux
  Techniques)
Byul. Acad. Nauk Belorussk.
  SSR
  Byulleten. Akademiya Nauk
  Belorusskoi SSR
Byull. Vses. N.-I. Inst. Kukuruzy
  Byulleten.' Vsesoyuznyi Nauchno-

Issledovatel'skii Institut
Kukuruzy
Cah. Agric. Prat. Pays Chauds
  Cahiers d'Agriculture Pratique
  des Pays Chauds
Cah. Rech. Agron.
  Cahiers de la Recherche
  Agronomique
Cahiers Nutrition Dietetique
  Cahiers de Nutrition et de
  Diététique
Can. Geogr. J.
  Canadian Geographical Journal
Can. J. Bot.
  Canadian Journal of Botany
Can. J. Pl. Sci.
  Canadian Journal of Plant Sci-
  ence
Can. J. Res. Sect. C. Bot. Sci.
  Canadian Journal of Research.
  Section C. Botanical Sci-
  ences
Canada Exp. Farm. Rep.
  Report on the Experimental
  Farms. Department of Ag-
  riculture, Canada
Cane Growers Quart. Bull.
  Cane Growers' Quarterly
  Bulletin
Caryologia
  Caryologia
Ceylon Adm. Rep. Agric.
  Ceylon Administration Reports.
  Agriculture
Ceylon J. Med. Sci.
  Ceylon Journal of Medical
  Science
Chin. J. Exp. Biol.
  Chinese Journal of Experi-
  mental Biology
Chin. J. Nutr.
  Chinese Journal of Nutrition
Chin. J. Sci. Agr.
  Chinese Journal of the Sci-
  ence of Agriculture
Chromosoma
  Chromosoma
Chung-kuo Nung-yeh Ko-hsueh
  Chungkuo Nungyeh Kohsueh
Cir. Dep. Agric. Uganda
  Circular of the Department of
  Agriculture, Uganda Pro-

tectorate
Collnea Bot., Barcinone
  Collectanea Botanica a Barci-
  nonensi Botanico Instituto
  Edita
Colo. Agric. Exp. Sta. Bull.
  Colorado. Agricultural Ex-
  periment Station. Bulletin
Colo. Sta. Press Bul.
  Colorado Agricultural Experi-
  ment Station. Press Bulle-
  tin
C. R. Acad. Sci. U. R. S. S.
  Compte Rendu de l'Academie
  des Sciences de l'U. R. S. S.
C. R. Hebd. Seanc. Acad. Sci.,
  Paris
  Compte Rendu Hebdomadaire
  des Seances de l'Academie
  des Sciences
Crop Sci.
  Crop Science
Crops Soils
  Crops and Soils
Curr. Sci.
  Current Science (India)
Cytologia
  Cytologia (Tokyo)

Diss. Abstr.
  Dissertation Abstract
Dnepropetrovsh.
  Dnepropetrovskiy Sel'skokhoz-
  yaystvennyy Institut. Trudy
Dokl. Akad. Nauk Armyansk. SSR
  Doklady Akademii Nauk
  Armyanskoi SSR
Dokl. Akad. Nauk SSSR
  Doklady Akademii Nauk SSSR
Dokl. Mosk. Sel'khoz. Akad.
  K. A. Timiryazeva
  Doklady Moskovskoi Sel'sko-
  khozyaistvennoi Akademii im.
  K.A. Timiryazeva
Dtsch. Landw.
  Deutsche Landwirtschaft
Dtsch. Landw. Pr.
  Deutsche Landwirt-schaftliche
  Presse

E. A. A. F. R. O. Annual Report
  East African Agriculture and
  Forestry Research Organi-
  zation. Annual Report

E. African Med. J.
    East African Medical Journal
East Afr. Agr. J.
    The East African Agricultural
    Journal
Ecology
    Ecology
Econ. Bot.
    Economic Botany
Emp. Cott. Gr. Rev.
    Empire Cotton Growing Review
Emp. J. Exp. Agric.
    The Empire Journal of Ex-
    perimental Agriculture
Ent. Mon. Mag.
    Entomologist's Monthly Maga-
    zine
Exp. Agr.
    Experimental Agriculture
Exp. Prog. Grassl. Res. Inst.
    Hurley
    Experiments in Progress.
    Grassland Research Insti-
    tute, Hurley
Exp. Sta. Rec.
    Experiment Station Record

FAO Plant Protect. Bull.
    FAO Plant Protection Bulletin
Feddes Rep.
    Feddes Repertorium Speci-
    erum Novarum Regni Vege-
    tablis
Fiziol. Rast.
    Fiziologiya Rastenii
Florida Sta. Bull
    Florida. Agricultural Experi-
    ment Station. Bulletin
Fmg. S. Afr.
    Farming in South Africa
Folia Quarternaria
    Folia Quarternaria
Forschungsdienst
    Der Forschungsdienst; organ
    der Deutschen Landbau-
    wissenschaft
Fortschr. Landw.
    Fortschritte der Landwirt-
    schaft

Ga. Agric. Exp. Sta. Ann. Rpt.
    Georgia. Agricultural Experi-
    ment Stations. Annual Re-
    port
Ga. Agric. Exp. Sta. Mimeo Ser.
    Georgia. Agricultural Experi-
    ment Station. Mimeograph
    Series N. S.
Ga. Agric. Res.
    Georgia Agricultural Research
Ga. Exp. Sta. Leaflet
    Georgia. Experiment Station.
    Leaflet
Genetica
    Genetica
Genetics
    Genetics
Gig. Sanit.
    Gigiena i Sanitariia
Grasas Aceites
    Grasas y Aceites
Grass Mimeograph
    Grass Mimeograph. Texas
    Research Foundation

Hawaii Sta. Rpt.
    Hawaii. Agricultural Experi-
    ment Station. Report
Herb. Absts.
    Herbage Abstracts
Heredity
    Heredity
Hindustan Antibiot. Bull.
    Hindustan Antibiotics Bulletin
Höfchenbr. Bayer Pflschutz-
    Nachr. Eng. Ed.
    Höfchen-Briefe. Bayer Pflan-
    zenschutz-Nachrichten Eng-
    lish Edition
Hsi-pei Nung-yeh Ko-hsueh
    Hsi-pei Nung Yeh K'o Hsüeh

Idengaku Zasshi
    see Japanese Journal of Gen-
    etics
Illinois Sta. Bul.
    Illinois. Agricultural Experi-
    ment Station. Bulletin
Indian Agr.
    Indian Agriculturist
Indian Cott. Gr. Rev.
    Indian Cotton Growing Review
Indian Fmg.
    Indian Farming
Indian J. Agric. Sci.
    Indian Journal of Agricultural
    Sci.

Indian J. Agron.
Indian Journal of Agronomy
Indian J. Ent.
Indian Journal of Entomology
Indian J. Genet. Pl. Breed.
Indian Journal of Genetics and
Plant Breeding
Indian J. Med. Res.
Indian Journal of Medical Re-
search
Indian J. Plant Physiol.
Indian Journal of Plant Physi-
ology
Indian J. Sci. Ind.
Indian Journal of Science and
Industry
Indian J. Sugarcane Res. Dev.
Indian Journal of Sugarcane
Research and Development
Indian Phytopathol.
Indian Phytopathology
Indian Sci. Abst.
Indian Science Abstracts
Indian Vet. J.
Indian Veterinary Journal
Indiana Sta. Rpt.
Indiana. Agricultural Experi-
ment Station. Annual Report
Int. Bull. Pl. Prot.
International Bulletin of Plant
Protection
Int. Congr. Soil Sci. Rep.
International Congress of Soil
Science. Report
Iowa St. Bul. P.
Iowa. Agricultural Experiment
Station. Bulletin
Israel J. Agric. Res.
Israel Journal of Agricultural
Research
Ital. Agric.
Italia Agricòla
Izv. Akad. Nauk. Kaz. SSR. Ser.
Biol.
Izvestiya Akademii Nauk Kaza-
khskoi SSR. Ser. Biol.
Izv. Kamar. Narod. Kult.
Izvestiya Na Kamarata Na
Narodnata Kultura
Izv. Kujbyšev. Sel'skohoz. Inst.
Izvestiya Kujbyshevskago Sel'-
skokhozyaistvennago Insti-
tuta

Izv. Moskov. Selsk. Khoz. Inst.
Izvestiya Moskovskago Sel'-
skokhozyaistvennago Institu-
ta
J. Agr. Food Chem.
Journal of Agricultural and
Food Chemistry
J. Agr. Res.
Journal of Agricultural Re-
search
J. Agric. Exp. Sta. Tokyo
Journal of the Imperial Agri-
cultural Experiment Station,
Nishigahara, Tokyo
J. Agric. Sci.
Journal of Agricultural Science
J. Agric. Trop. Bot. Appl.
Journal d'Agriculture Tropi-
cale et de Botanique Appli-
quee
J. Agric. Univ. Puerto Rico
Journal of Agriculture of the
University of Puerto Rico
J. Amer. Chem. Soc.
Journal of the American Chem-
ical Society
J. Amer. Soc. Agron.
Journal of the American So-
ciety of Agronomy
J. Amer. Soc. Sugar Beet Tech-
nol.
Journal of the American So-
ciety of Sugar Beet Tech-
nologists
J. Aust. Inst. Agric. Sci.
Journal of the Australian Insti-
tute of Agricultural Science
J. Biol. Chem.
Journal of Biological Chemis-
try
J. Dairy Sci.
Journal of Dairy Science
J. Dent. Res.
Journal of Dental Research
J. Dep. Agri. West. Aust.
Journal of the Department of
Agriculture for Western
Australia
J. Econ. Entomol.
Journal of Economic Entomol-
ogy
J. Exp. Landw. Sudosten. Eur.
Russl.

Journal für experimentelle
Landwirtschaft im Südosten
des Eur.-Russlands
J. Fac. Agric. Hokkaido Univ.
Journal of the Faculty of Ag-
riculture, Hokkaido Univer-
sity
J. Fac. Sci. Tokyo Univ. : Sect.
III
Journal of the Faculty of Sci-
ence, Tokyo University.
Section III. Botany
J. Genetics
Journal of Genetics
J. Hered.
Journal of Heredity
J. Ind. Engng. Chem.
Journal of Industrial and Engi-
neering Chemistry
J. Indian Bot. Soc.
Journal of the Indian Botani-
cal Society
J. Indian Inst. Sci.
Journal of the Indian Institute
of Science
J. Indian Soc. Soil Sci.
Journal of the Indian Society
of Soil Science
J. Jap. Bot.
Journal of Japanese Botany
J. Jap. Soc. Food Nutr.
Journal of the Japanese Society
of Food and Nutrition
J. Jap. Soc. Grassland Sci.
Journal of the Japanese Society
of Grassland Science
J. Madras Univ. B.
Journal of Madras University.
Section B. Contributions in
Mathematics, Physical and
Biological Sciences
J. Mysore Agric. Exp. Union
Journal of the Mysore Agricul-
tural and Experimental Union
J. Nutr. Diet.
Journal of Nutrition and Die-
tetics
J. Plant Prot.
Journal of Plant Protection,
Tokyo
J. Postgrad. Med.
Journal of Postgraduate Medi-
cine

J. Postgrad. Sch. , IARI
Journal of the Post-Graduate
School, Indian Agricultural
Research Institute, New Del-
hi
J. Res. (India)
Journal of Research, Punjab
Agricultural University,
Ludhiana
J. Sci. Food Agric.
Journal of the Science of Food
and Agriculture
J. Sci. Industr. Res.
Journal of Scientific and Indus-
trial Research
J. Soil Wat. Conserv.
Journal of Soil and Water Con-
servation
J. W. Afr. Inst. Oil Pal Res.
Journal of the West African
Institute for Oil Palm Re-
search
J. Wash. Acad. Sci.
Journal of the Washington
Academy of Sciences
Jap. J. Bot.
Japanese Journal of Botany
Jap. J. Genet.
Japanese Journal of Genetics

Kew Bull.
Kew Bulletin
Khimiya Sel'khoz
Khimiya v Sel'skom Khozya-
istve
Kirkia
Kirkia
Kolkhoz. Proizv.
Kolkhoznoe Proizvodstvo
Korte Ber.
Korte Berichten
Krishik
Krishik
Kulturpflanze
Kulturpflanze
Kumamoto J. Sci. Ser. "B" Sec. 2.
Biol.
Kumamoto Journal of Science.
Series B. Section 2. Biol-
ogy

LABDEV
LABDEV

Mycol. Cir. Dep. Agric. Tanga-
  nyika
  Mycological Circular. Depart-
    ment of Agriculture, Tanga-
    nyika
Mycol. Pap.
  Mycological Papers. Common-
    wealth Mycological Institute,
    Kew
Mycologia
  Mycologia
Mycopathol. Mycol. Appl.
  Mycopathologia et Mycologia
    Applicata
Mysore Agric. Cal. Yb.
  Mysore Agricultural Calendar
    and Yearbook
Mysore Agric. J.
  Mysore Agricultural Journal
Mysore Dep. Agr. Rpt.
  Report of the Department of
    Agriculture, Mysore
Mysore J. Agric. Sci.
  Mysore Journal of Agricultur-
    al Sciences

N. Dak. Agr. Exp. Sta. Bul.
  North Dakota. Agricultural Ex-
    periment Station. Bulletin
N. Dak. Farm Res.
  North Dakota Farm Research
N. Dak. Sta. Bimo. Bul.
  North Dakota. Agricultural
    Experiment Station. Bi-
    monthly Bulletin
N. Dak. Sta. Bull.
  North Dakota. Agricultural
    Experiment Station. Bulle-
    tin
N. Dak. Sta. Rpt.
  North Dakota. Agricultural Ex-
    periment Station. Report
Nachr. Schadlebekampf.
  Nachrichten über Schädlings-
    bekämpfung
Nachrbl. Dtsch. Pflschdienst
  Nachrichtenblatt für den
    Deutschen Pflanzenschutz-
    dienst
Nagpur Agric. Coll. Mag.
  Nagpur Agricultural College
    Magazine
Nat. Inst. Anim. Ind. (Chiba) Bull.

National Institute of Animal
    Industry, Chiba. Bulletin
National Seeds Corporation Ltd.
    Bulletin
  National Seeds Corporation,
    Ltd. Bulletin
Nature
  Nature (London)
Nauch. Dokl. Vyssh. Shk. Biol.
    Nauk
  Nauchnye Doklady Vysshei
    Shkoly., Biol. Nauk
Nauch. Sotrud. Vses. Inst. Ras-
    tenievod.
  Nauchnyk Sotrudnikov. Vsesoy-
    uznyi Institut Rastenievod-
    stva
Nauk. Pered. Opyt Sel' Khoz.
  Nauka i Peredovoi. Opyt v
    Sel'skom Khozyaistve
Nauk. Prac. Nauk-Doslid. Inst.
    Zemlerob. Tvarynn Zakhid
    Rai. URSR
  Naukovi Pratsi. Naukovo-
    Doslidny Institut Zemlerob-
    stva i Tvarynnystva Zakhid-
    nykh Raicniv URSR
Nebr. Agr. Exp. Sta. Bull.
  Nebraska. Agricultural Experi-
    ment Station. Bulletin
Nematologica
  Nematologica
New Hampshire Sta. Bul.
  New Hampshire. Agricultural
    Experiment Station. Bulle-
    tin
New Jersey Sta. Bul.
  New Jersey. Agricultural Ex-
    periment Station. Bulletin
New Jersey Sta. Rpt.
  New Jersey. Agricultural Ex-
    periment Station. Annual
    Report
New Mexico St. Bul.
  New Mexico. Agricultural Ex-
    periment Station. Bulletin
New Zealand J. Agr.
  New Zealand Journal of Agri-
    culture
Notes from Agric. Exp. Sta.
    Korea
  Notes from Agricultural Ex-
    periment Stations, Korea

Novenytermeles
Novenytermeles
Nuclear Sci. Abst.
Nuclear Science Abstracts
Nuovo Gior. Bot. Ital.
Nuovo Giornale Botanico Ital-
iano
Nutritio et Dieta
Nutritio et Dieta
Nyasaland Agric. Quart. J.
Nyasaland Agricultural Quar-
terly Journal

Ogiz. Selkolkhozguz, Moscow
Ogiz. Selkolkhozgus, Moscow
Ohio St. Bul.
Ohio. Agricultural Experiment
Station. Bulletin
Oleagineux
Oleagineux
Opyt. Agron.
Opytnaya Agronomiya
Oregon Sta. Circ.
Oregon. Agricultural Experi-
ment Station. Circular of
Information

Pam. Panst. Inst. Nauk. Gosp.
Wiejsk. Pulawach
Pamietnik Panstwowego Insty-
tutu Naukowego Gospodarst-
wa Wiejskiego Pulawach
Pam. Pulawski
Pamietnik Pulawski
Pamph. Djetyssouy Pl. Prot. Sta.
Alma-Ata.
Pamphlet Djetyssouy Plant
Protection Station, Alma-Ata
Pflanzenschutzberichte
Pflanzenschutzberichte
Philipp. Agric.
Philippine Agriculturist
Philipp. Agric. Rev.
Philippine Agricultural Review
Philipp. J. Agric.
Philippine Journal of Agricul-
ture
Philipp. J. Pl. Indust.
Philippine Journal of Plant In-
dustry
Physiol. Plant.
Physiologia Plantarum
Phytomorphology

Phytomorphology
Phyton
Phyton
Phytopath. Z.
Phytopathologische Zeitschrift
Phytopathology
Phytopathology
Pl. Dis. Reptr.
Plant Disease Reporter
Pl. Soil
Plant and Soil
Plant Food
Plant Food Review
Plant Int. Rev.
Plant Introduction Review
Plant Prot. Insect. Dis.
Zashchita Rastenii of Vredi-
telei i Boleznei
Pochvovedenie
Pochvovedenie
Polish Agric. & Forest Ann.
Polish Agricultural and Forest
Annual
Poona Agric. Coll. Mag.
Poona Agricultural College
Magazine
Prikl. Biokhim. Mikrobiol.
Prikladnaya Biokhimiia i Mik-
robiologiia
Proc. Amer. Soc. Hort. Sci.
Proceedings. American Society
for Horticultural Science
Proc. Ass. Seed Anal. N. Amer.
Proceedings of the Official
Seed Analysts of North
America
Proc. Ass. So. Agric. Wkrs.
Proceedings of the Association
of Southern Agricultural
Workers
Proc. Bihar Acad. Agric. Sci.
Proceedings of the Bihar Acad-
emy of Agricultural Sciences
Proc. Crop Sci. Soc. Japan
Proceedings of the Crop Sci-
ence Society of Japan
Proc. Indian Acad. Sci. Sect. B.
Proceedings of the Indian
Academy of Sciences. Sec-
tion B
Proc. Indian Sci. Congr.
Proceedings of the Indian Sci-
ence Congress

Proc. Int. Bot. Cong.
Proceedings of the International Botanical Congress
Proc. Int. Conf. Genet.
Proceedings of the International Conference on Genetics
Proc. Int. Seed Test. Ass.
Proceedings of the International Seed Testing Association
Proc. Lenin Acad. Agric. Sci.
Doklady Vsesoyuznoi Akademii Sel'skokhozyaistvennykh Nauk im. V. I. Lenina
Proc. Linn. Soc. London.
Proceedings of the Linnaean Society of London
Proc. Nat. Acad. Sci. India
Proceedings of the National Academy of Sciences of India
Proc. Nat. Acad. Sci. Wash.
Proceedings of the National Academy of Sciences of the U. S. A.
Proc. Nat. Inst. Sci. India (B)
Proceedings of the National Institute of Sciences of India. Ser. B. Biological Sciences
Proc. Pakistan Sci. Conf.
Proceedings of the Pakistan Science Conference
Proc. Sci. Wkrs. Conf., Coimbatore
Proceedings of the Scientific Workers' Conference - Coimbatore
Proc. Soil Crop. Sci. Soc. Fla.
Proceedings of the Soil and Crop Sciences Society of Florida
Proc. Tartar Agric. Res. Sta.
Proceedings of the Tartar Agricultural Research Station
Prog. Rep. Pa. Agric. Exp. Sta.
Progress Report. Pennsylvania Agricultural Experiment Station
Progressive Farmer (Georgia-Alabama-Florida edition)
Progressive Farmer (Georgia-Alabama-Florida edition)
Psychopharmacologia
Psychopharmacologia

Pubbl. Comitato Sci. Aliment. R. Accad. Naz. Lincei
Pubblicazioni. Comitato Scientifico per l'Alimentazione. R. Accademia Nazionale dei Lincei
Publ. Inst. Nat. Agron. Congo Belge
Publications de l'Institut National pour l'Etude Agronomique du Congo Belge
Pustyni, U. S. S. R. i ikh Osvoenie
Pustyni, U. S. S. R. i ikh Osvoenie

Qd. Agric. J.
Queensland Agricultural Journal
Qd. Dep. Agr. Stock Ann. Rpt.
Queensland. Department of Agriculture and Stock. Annual Report
Qual. Plant Mater. Veg.
Qualitas Plantarum et Materiae Vegetabiles

Rapp. Minist. Agric. Prov. Queb.
Rapport. Ministere de l'agricultura de la Province de Quebec
Rast. Nauki, Sofia
Rastenievadni Nauki, Sofia
Referat. Zh.
Referativnyi Zhurnal
Referat. Zh. Biol.
Referativnyi Zhurnal, Biologiya
Referat. Zh. Rasten.
Referativnyi Zhurnal. Otdel'nyi vypusk: Rastenievodstvo
Rend. Ist. Superiore Sanita (Rome)
Rendiconti dell'Istituto Superiore di Sanita. Roma
Rep. Ariz. Agric. Exp. Sta.
Report. Arizona Agricultural Experiment Station
Rep. Bd. Sci. Adv. India
Report of the Board of Scientific Advice for India
Rep. Dep. Agric. Bombay.
Report. Department of Agri-

culture, Bombay
Rep. Dep. Agric. Cent. Prov.
Berar
Report of the Department of
Agriculture, Central Prov-
inces and Berar
Rep. Dep. Agric. Kenya
Report. Department of Agri-
culture, Kenya
Rep. Dep. Agric. Nyasald.
Report. Department of Agri-
culture, Nyasaland
Rep. Dep. Agric. Tanganyika
Report. Department of Agri-
culture, Tanganyika
Rep. Dep. Agric. Uganda
Report of the Department of
Agriculture, Uganda
Rep. Div. Ld. Res. Reg. Surv.
C. S. I. R. O. Aust.
Report. Division of Land Re-
search and Regional Survey,
C. S. I. R. O., Australia
Rep. Div. Pl. Ind. C. J. I. R. O.
Aust.
Report. Division of Plant In-
dustry, C. S. I. R. O., Aus-
tralia
Rep. Div. Trop. Past., C. S. I.
R. O. Aust.
Report. Division of Tropical
Pastures, C. S. I. R. O., Aus-
tralia
Rep. Fla. Agric. Exp. Sta.
Report of the Florida Agricul-
tural Experiment Station
Rep. Grassl. Res. Sta. Maran-
dellas
Report. Grasslands Agricultur-
al Research Station, Maran-
dellas, Southern Rhodesia
Rep. Indian Coun. Agric. Res.
Report. Indian Council of Ag-
ricultural Research
Rep. Kihara Inst. Biol. Res.
Report of the Kihara Institute
for Biological Research
Res. and Indry.
Research & Industry (New Del-
hi)
Res. Bull. Hokkaido Nat. Agr.
Exp. Sta.
Research Bulletin. Hakkaido

National Agricultural Ex-
periment Station
Res. Bull. Ia. Agric. Exp. Sta.
Research Bulletin. Iowa Agri-
cultural Experiment Station
Res. Bull. S. Manchurian Rly.
Co.
Research Bulletin of the Agri-
cultural Experiment Station
of the South Manchurian
Railway Company
Res. Mem. Emp. Cott. Gr. Corp.
Research Memoirs. Empire
Cotton Growing Corporation
Res. Progr. Rep. Purdue Univ.
Research Progress Report.
Purdue University
Res. Rep. N. Cent. Weed. Contr.
Conf.
Research Report. North Cen-
tral Weed Control Confer-
ence
Res. Ser. Indian Coun. Agric.
Res.
Research Series. Indian Coun-
cil of Agricultural Research
Rev. Agric. Sucr. Maurice
Revue Agricole et Sucriere de
l'Ile Maurice
Rev. Appl. Mycol.
Review of Applied Mycology
Rev. Argent. Agron.
Revista Argentina de Agrono-
mia
Rev. Bot. Appl.
Revue de Botanique Appliquee
et d'Agriculture Tropicale
Rev. Inst. Recherches Agron.
Bulg.
Revue des Instituts de Re-
cherches Agronomiques en
Bulgarie
Rev. Nutricao
Revista de Nutrição
Rev. Path. Veg.
Revue de Pathologie Vegetale
et d'Entomologie Agricole
de France
Revta Invest. Agropec.
Revista de Investigaciones
Agropecuarias
Rhod. Agric. J.
Rhodesia Agricultural Journal

Rhode Island Sta. Bul.
  Rhode Island. Agricultural
  Experiment Station. Bulle-
  tin
Rice J.
  Rice Journal
Riv. Agric. Subtrop. Trop.
  Rivista di Agricoltura Sub-
  tropicale e Tropicale
Riz et Rizicult.
  Riz et Riziculture
Rpt. Storrs Agric. Exp. Sta.
  Report of the Storrs Agricul-
  tural Experiment Station,
  Storrs, Conn.

S. Afr. J. Agric. Sci.
  South African Journal of Ag-
  ricultural Science
S. Carolina Agric. Exp. Sta.
  Bull.
  South Carolina. Agricultural
  Experiment Station. Bulle-
  tin
S. Dak. Acad. Sci. Proc.
  South Dakota Academy of Sci-
  ence. Proceedings
S. Dak. Agric. Exp. Sta. Bull.
  South Dakota. Agricultural
  Experiment Station. Bulle-
  tin
S. Dak. Fm. Home Res.
  South Dakota Farm and Home
  Research
S. Dak. Sta. Bul.
  South Dakota. Agricultural Ex-
  periment Station. Bulletin
Saatgut-Wirt.
  Saatgut-Wirtschaft
Samaru Res. Bull.
  Samaru Research Bulletin
Sb. Csl. Acad. Zemed.
  Sbornik Ceskoslovenske Aka-
  demie Zemedelske
Sb. Nauc. -Issled. Rab. Stud.
  Stavropol'sk. S-Kh. Inst.
  Sbornik Nauchno-Issledovatel'
  Skikh Rabot Studentov Stav-
  ropol'skoga Sel'skokhozy-
  aistvennogo Instituta
Sb. Tr. Aspirantov Molodykh
  Nauch. Sotrudnikov Vses. Inst.
  Rastevodstva.

Sbornik Trudov Aspirantov i
  Molodykh Nauchnykh Sotrud-
  nikov. Vsesoyuznyi Insti-
  tut Rastenievodstva
Sci. Abstr. China Biol. Sci.
  Science Abstracts of China.
  Biological Sciences
Sci. Bull. Dep. Agric. S. Afr.
  Scientific Bulletin. Department
  of Agriculture, South Af-
  rica
Sci. Cult.
  Science and Culture
Sci. Monogr. Indian Coun. Agric.
  Res.
  Scientific Monograph. Indian
  Council of Agricultural Re-
  search
Sci. Rec. Acad. Sin.
  Science Record. Academia
  Sinica
Sci. Rep. Agric. Res. Inst. Pusa
  Scientific Reports of the Agri-
  cultural Research Institute,
  Pusa
Sci. Rep. Imp. Inst. Agric.
  Res. Pusa
  Scientific Reports of the Im-
  perial Institute of Agricul-
  tural Research, Pusa
Sci. Rep. Indian Agric. Res.
  Inst.
  Scientific Reports of the Indi-
  an Agricultural Research
  Institute, New Delhi
Sci. Rep. Kyoto Prefect. Univ.,
  Agric.
  Scientific Reports of the Ky-
  oto Prefectural University.
  Agriculture.
Sci. Rep. Saikyo Univ., Agric.
  Scientific Reports of the Sai-
  kyo University. Agriculture
Sci. Rep. Sch. Biol. Sci.
  Scientific Reports of the
  School of Biological Sci-
  ences
Science
  Science
Scientist (Pakistan)
  Scientist (Pakistan)
Seed World
  Seed World

Selek. Semenovodstvo
Selektsiya i Semenovodstvo
Sel'hoz. Sibiri
Sel'skoe Khozyaistvo Sibiri
Selsk. Khoz. i Lesov.
Sel'skoe Khozyaistvo i
Lesovodstvo
Selskostop. Nauka
Selskostopanska Nauka
Sementi Elette
Sementi Elette
Shokubutsu Kennyu Zasshi
see Journal of Japanese
Botany
S. -Kh. Inst.
Stavropol'ski Nauchno-Issledo-
vatel'skii Instituti Sel'skogo
Khozyaistva. Trudy
Soc. Zern. Hoz.
Sotsialisticheskoe Zernovoe
Khozyaistvo
Soil Sci. Soc. Amer. Proc.
Soil Science of America.
Proceedings
Sols Afr.
Sols Africains
Sorghum Newsletter
Sorghum Newsletter
South Car. Sta. Bull.
South Carolina. Agricultural
Experiment Station. Bulletin
Soviet Plant Industry Record
Vestnik Sotsialisticheskogo
Rastenievodstva
Sth. Seedsman
Southern Seedsman
Span
Span
Spirto-Vodochnaya Prom.
Spirtovaya-Vodochnaya Prom-
yshlennost'
Sta. Circ. Wash. Agric. Exp.
Sta.
Station Circular. Washington
Agricultural Experiment
Station, Pullman.
Sudan Notes and Records
Sudan Notes and Records
Summ. Sci. Wk. Inst. Pl. Prot.
Leningrad
Summary of the Scientific
Work of the Institute for
Plant Protection, Leningrad

Sveriges Utsadesfor. Tidskr.
Sveriges Utsädesförenings Tids-
krift
Symp. Sci. Wk. Sci.-Res. Inst.
Agric. Anim. Husband. W.
Distr. Ukrain. S. S. R.
Symposium on Scientific Work.
Scientific-Research Institute
of Agriculture and Animal
Husbandry, Western Dis-
trict, Ukrainian S. S. R.
Tech. Bull. Colo. Agric. Exp.
Station
Technical Bulletin. Colorado
Agricultural Experiment
Station
Tech. Pap. Div. Ld. Res. Reg.
Surv. C. S. I. R. O.
Technical Papers. Division of
Land Research and Region-
al Survey, C. S. I. R. O.,
Australia
Tenn. Farm Home Sci. Prog.
Rep.
Tennessee Farm and Home
Science Progress Report
Teysmannia
Teysmannia
Tijdschr. Planteziekten
Tijdschrift over Plantenziekten
Tr. Bot. Inst. Akad. Nauk SSSR
Trudy Botanicheskogo Insti-
tuta. Akademiya Nauk SSSR
Tr. Bot. Inst. V. L. Komarova
Trudy Botanicheskii Institut
V. L. Komarova
Tr. Byuro Prikl. Bot.
Trudy Byuro po Prikladnoi
Botanike
Tr. Gruzinsk. Sel'skokhoz. Inst.
Trudy Gruzinskogo Sel'skok-
hozyaistvennogo Instituta im
L. P. Beriya
Tr. Inst. Bot. V. L. Komarova
Trudy. Institut Botaniki im V.
L. Komarova
Tr. Inst. Fiziol. Rast. im. K. A.
Timirjazeva
Trudy Instituta Fiziologii Ras-
tenii im. K. A. Timiryazeva
Tr. Inst. Genet. Akad. Nauk
SSSR

Trudy Instituta Genetiki. Aka-
  demiya Nauk SSR
Tr. Inst. Genet. i Selektsii
Trudy Instituta Genetiki i Sel-
  ektsii. Akademiya Nauk
  Azerbaidzhanskoi SSR
Tr. Khar'kovsk. Sel'skokhoz.
  Inst.
Trudy Khar'kovskogo Sel'sko-
  khozyaistvennogo Instituta
Tr. Mosk. Obshch. Ispyt. Prir.
Trudy Moskovskogo Obshchest-
  va Ispytatelei Prirody
Tr. Prik. Bot. Genet. Selek.
Trudy po Prikladnoi Botanike,
  Genetike i Selektsii
Tr. Prik. Bot. i. Selek.
Trudy po Prikladnoi Botanike
  i Selektsii
Tr. Sel'. -Kohz. Met.
Trudy po Sel'skokhozyaistven-
  noi Meteorologii
Tr. Tsent. Inst. Prognozov
Trudy Tsentral'nogo Instituta
  Prognozov
Tr. Ukrain. Nauch. -Issledovatel.
  Inst. Rastenievodstva, Selekt-
  sii i Genet.
Trudy. Ukrainskii Nuchno-
  Issledovatel'skii Institut
  Rastenievodstva, Selektsii i
  Genetiki
Tr. Volgograd Sel'. -Khoz. Inst.
Trudy. Volgograd Sel'skokhoz-
  yaistvennyi Institut
Tr. Vses. Nauchb. -Issled. Inst.
  Udobrenii i Agropochvoved-
  eniya
Trudy Vsesoyuznogo Nauchno-
  Issledovatelskogo Instituta
  Udobrenii, Agrotekhniki i
  Agropochvovedeniya
Trans. Brit. Mycol. Soc.
Transactions of the British
  Mycological Society
Trans. Jn. Sci. Sci. -Res. Inst.
  Univ. Min. Agric. Uzbec SSR
Transactions Journal of Sci-
  ence, Scientific-Research
  Institute. University, Min-
  istry of Agriculture, Uzbec
  SSR
Trans. Orenburg Agric. Inst.

Transactions of the Orenburg
  Agricultural Institute imeni
  A. A. Andreyeva
Trans. R. Ent. Soc. Lond.
Transactions of the Royal
  Entomological Society of
  London
Trans. Roy. Soc. N. Z.
Transactions of the Royal So-
  ciety of New Zealand
Trans. Sapporo Nat. Hist. Soc.
Transactions of the Sapporo
  Natural History Society
Trans. Sci. -Res. Inst. Agric.
  Cent. Dist. Non-Chernozem
  Zone
Transactions, Scientific-Re-
  search Institute of Agricul-
  ture. Central District, Non-
  Chernozem Zone
Trans. Sci. -Res. Inst. Fertiliz.
  Agric. Soil Sci.
Transactions. All-Union Scien-
  tific-Research Institute of
  Fertilizers and Soil Science
Trop. Agric. Trin.
Tropical Agriculture (Trini-
  dad)
Trop. Agriculturist
Tropical Agriculturist and
  Magazine of the Ceylon Ag-
  ricultural Society
Trop. Sci.
Tropical Science

Uch. Zap. Kuybyshevsk. Gos.
  Ped. Inst.
Uchenye Zapiski. Kuybyshev.
  Gosudarstvennyi Institut
Uchenye Zap. Kharkov. Univ.
Uchenye Zapiski Khar'kovskogo
  Universiteta
Ukr. Bot. Zh.
Ukrayins'kyi Botanichnyi Zhur-
  nal
Univ. Ankara Fac. Agr. Year-
  book
University of Ankara. Faculty
  of Agriculture. Yearbook
Univ. of Fla. Agr. Exp. Sta.
  Ann. Report
University of Florida. Agricul-
  tural Experiment Station.
  Annual Report

Univ. of Nebr. Quarterly
  Nebraska Experiment Station
  Quarterly
USDA Agr. Bull.
  U.S. Department of Agricul-
  ture. Agriculture Informa-
  tion Bulletin
USDA Bull.
  U.S. Department of Agricul-
  ture. Bulletin
USDA Bur. Pl. Indus. Bull.
  U.S. Department of Agricul-
  ture. Bureau of Plant In-
  dustry. Bulletin
USDA Bur. Pl. Indus., Work
  Yuma Exp. Farm
  U.S. Department of Agricul-
  ture. Bureau of Plant In-
  dustry. Work of the Yuma
  Experimental Farm
USDA Div. Ent. Bul.
  U.S. Department of Agricul-
  ture. Division of Entomol
  ogy. Bulletin
USDA Farmers Bull.
  U.S. Department of Agricul-
  ture. Farmers Bulletin
USDA Leaflet
  U.S. Department of Agricul-
  ture. Leaflet
USDA Misc. Publ.
  U.S. Department of Agricul-
  ture. Miscellaneous publi-
  cation
Ust. Ved. Inf. MZLVH Rostl.
  Vyroba
  Czechoslovakia. Ministerstvo
  Zemedelstvi, Lesniho a
  Vodniho Hospodarstvi. Ustav
  Vedeckotechnickych Infor-
  maci. Rostlinna Vyroba

Var. Rep. Alabama Agric. Exp.
  Sta.
  Variety Reports. Alabama Ag-
  ricultural Experiment Sta-
  tion
Ved. Pr. Vyzk. Ustavu Rostlinne
  Vyroby Piestanoch
  Vedecke Prace Vyzkumneho
  Ustavu Rostinne Vyroby
  Piestany
Vermont St. Bul.

Vermont Agricultural Experi-
  ment Station. Bulletin
Vest. Moskov. Univ. Ser. VI.
  Biol. Pochvoved.
  Vestnik Moskovskogo Univer-
  siteta. Ser. 6: Biologiya,
  Pochvovedenie
Vestn. Sel-Kh. Nauk.
  Vestnik Sel'skokhozyaistvennoi
  Nauki
Veterinariya
  Veterinariya
Virginia Sta. Bul.
  Virginia. Agricultural Experi-
  ment Station. Bulletin
Vop. Virusl.
  Voprosy Virusologii
Vrtljschr. Bayer. Landw. Rat.
  Vierteljahrsschrift des Bayer-
  ischen Landwirtschaftsrats

W. Pakistan J. Agric. Res.
  West Pakistan Journal of Ag-
  ricultural Research
Weed Res.
  Weed Research
Weed Sci.
  Weed Science
Weeds
  Weeds
Wld. Fmg.
  World Farming

Z. Bot.
  Zeitschrift für Botanik
Z. Pflanzenz
  Zeitschrift für Pflanzenzücht-
  tung
Zap. Leningradsk. Sel'skokhoz.
  Inst.
  Zapiski Leningradskogo Sel'sko-
  khozyaistvennogo Instituta
Zashch. Rast. Vredit. Bolez.
  Zashchita Rastenii ot Vredi-
  telei i Boleznei
Zast. Bilja
  Zastita Bilja
Zbl. Bakt., Abt. 2
  Zentralbaltt für Bakteriologie,
  Parasitenkunde, Infektions-
  krankheiten und Hygiene.
  Abt. 2

Zemledelie
  Zemledelie
Zh. Opyt. Agron.
  Zhurnal Opytnoi Agronomii
Zhivotnovodstvo
  Zhivotnovodstvo
Zischr. Pflanzenkrank.
  Zeitschrift für Pflanzenkrank-
    heiten und Pflanzenschutz
Ztschr. Landw. Versuchsw.
  Osterr.
  Zeitschrift für das Landwirt-
    schaftliche Versuchwesen in
    Österreich

MILLETS BIBLIOGRAPHY

1 ABICHANDI, C. T. , and BHATT, P. N.
Salt tolerance of germination of bajra (Pennisetum typhoides) and jowar (Sorghum vulgare) varieties. Ann. Arid Zone 1965:4 (1) 36-42.

2 ABIFARIN, A. O.
Research programme on Pennisetum millets at Kano, Northern Nigeria. Sols Afr. 11(1 and 2):139-142. 1966.

3 ADEYEMI, A. O.
Storage pests of cereals (maize, millet, Guinea corn) and their control in Nigeria. Sols Afr. 1967, 12, No. 2-3, 159-166 (Bibl. 21; Res. Div. Min. Agric. Nat. Resources, Ibadan, W. Nigeria).

4 ADRIAN, J. , FRANGNE, R. , BOULENGER, P. , DAVIN, A. , GALLANT, D. , ABEL, H. , GUILBOT, A. and GAST, M.
Bilan et interet nutritionnel de differents procedes de mouture du mil (Comparative value and nutritional interest of different methods of grinding millet). Cahiers Nutrition Dietetique 2(1):67-77. 1967.

5 ADRIAN, J. and JACQUOT, R.
Le Sorgho et les Mils en Alimentation Humaine et Animale. Vigot Freres, Editeurs. Paris 1964. pp. 189.

6 ADRIAN, J. , FRANGNE, R. , DAVIN, A. , GALLANT, D. and GAST, M.
The problem of the grinding of millet. Nutritional interest of methods of African artisans, and a trial of mechanisation. Agron. Trop. 1967, 22, 687-698.

7 ADRIANOV, V. S.
The effectiveness of granulated superphosphate. Zemledelie 1955 (3):77-78; Referat. Zh. Biol. 1956:1110. (BA 31:5656).

8 AFAFONOV, N. P.
Use of zinc for millet on typical chernozems of the Kursk Region. (Pochvovedenie 1964, No. 12:61-67. (SFA 28:1211).

9 AFANAS'EVA, A. S.
Autotetraploids of millet induced by colchicine. (pp. 154-

163 Polyploidy in Plants. Transactions of the Conference on Plant Polyploidy, 25-28 June 1958.) Ir. Mosk. Obschch. Ispyt. Prir. 1962:5:pp. 376 (Russian) [PBA 33:4195]

10 AFZAL, M.
    Moisture relations between soil and plant at the permanent wilting stage. W. Pakistan J. Agric. Res. 1:64-80. Ayub Agric. Res. Inst., Lyallpur. [SFA 27(4):2078].

11 AGARWAL, P. N. and SINHA, N. S.
    Chemical composition of the oil from the seeds of Pennisetum typhoideum (bajra). Indian J. Agron. 1964, 9, No. 4. 288-291.

12 AGARWALA, O. N., NEGI, S. S. and MAHADEVAN, V.
    Studies on the toxicity and nutritive value of fungus-free Paspalum scrobiculatum grains. Indian Vet. J. 1964, 41: No. 1:43-47.

13 AGATI, J. A. and CALICA, C.
    Studies on the host range of the rice and corn leaf-gall virus. Philipp. J. Agric. 15:249-257. 1950.

14 AGNIHOTRI, B. S. and BHIDE, V. P.
    Leaf-spot of sugarcane caused by Curvularia lunata (Wakker) Boedijn in Maharashtra State. Indian J. Sugarcane Res. Dev. 7(1):36-40. 1962.

15 AGRIC. GAZ. OF NEW SOUTH WALES
    Fodder millets, points in grazing. Agric. Gaz. N. S. W. 56. 52. 1945.

16 AGRICULTURAL RESEARCH, WASHINGTON
    Dwarf millet for high gains. Agric. Res. (Wash.) 16:11. 1968.

17 AGRIC. RESEARCH, WASHINGTON
    Pearl millet has twine. Agric. Res. (Wash.) 17(3):3. 1968.

18 AGURELL, S. and RAMSTAD, E.
    Analysis of Pennisetum ergot. Lloydia 25(2):67-77. 1962.

19 AHLRICH, V. E., YOUNG, W. C. and BYRD, M.
    New millet for wild life food. Miss. Fm. Res. 30:1, 3. Nov. 1967.

20 AHLUWALIA, M.
    A worthy successor to "Improved Ghana" bajra, Pusa Moti, Indian Fmg. 1964. 14:4-5.

21 AHLUWALIA, M. and SHANKAR, K.
    For better results with Improved Ghana. Indian Fmg. 1962,
    12: No. 7 p. 8.

22 AHLUWALIA, M. and SHANKAR, K.
    Inheritance of bristling in pearl millet.   Sci. Cult. 1964.
    30:340-341.

23 AHLUWALIA, M. and VITTAL RAO, D.
    Pusa Moti Bajra shows better adaptability in Andhra Pradesh.
    Andhra Agric. J. 1964, 11:160-161.

24 AIYADURAI, S. G. and RATNASWAMY, M. C.
    Kudiraivali, the barnyard millet (Echinchloa frumentacea
    Link) protects groundnut crop from the red hairy caterpillar.
    Madras Agr. J. 48(2): 60. 1961.

25 AJREKAR, S. L.
    Observations on a disease of jowar (Sorghum vulgare) caused
    by Sphacelia (conidial stage of Claviceps). J. Indian Bot.
    Soc. 2: 55-61. 1926.

26 AKMAN, ARIF VELI and PAMIR, M. HILMI
    On the microbiology of the Turkish Boza. Univ. Ankara Fac.
    Agr. Yearbook 33. 38 Illus. 1962. (Landw. Fak. Univ.
    Ankara, Ankara, Turk.)

27 ALEKSEEVA, E. S.
    Selection of paternal forms for isolation of varieties of Ital-
    ian millet. Agrobiologia 3:383-385. 1960. Referat. Zh.
    Biol., 1961 No. 5G353. (BA 36:65796).

28 AL-FAKHRY, A. K., GROGAN, C. O. and SARVELLA, P.
    Some genetic studies on babala, Pennisetum typhoideum L.
    (Rich.). Z. Pflanzenz 1965, 54: 182-187.

29 ALKAMPER, J.
    Spraying trials with simazine on various millet species.
    Bodenkultur 1966, 17(2): 174-186.

30 ALKAMPER, J.
    Simazine and atrazine spraying trials on various millet spe-
    cies. Bodenkultur 1966, 17, No. 4, 360-367.

31 ALKIPEROV, R. A.
    Weed control in millet sowing. Material Y Nauch. Konfer-
    entsii Izhevskii Sel'skokhoz. Inst. 9: 34-37. 1961; Referat.
    Zh. Biol., 1962, No. 10G485. (BA 42:3239).

32 AMES, J.W. and BOLTZ, G. E.
    Sulfur in relation to soils and crops. Ohio St. Bul. 292:221-
    256. 1961.

33 ANANJAN, V. L.
The effect of the radio-activity of soil air on plants. Doklady Akad. Nauk Armyansk. SSR 1962, 34: No. 3:113-116.

34 ANANTHAPADMANABHAN, C. D., APPADURAI, R. and SESHU, K. A.
Optimum dose of fertilisers for ragi in the lower Bhadvani project. Madras Agr. J. 1967. 54. No. 12 609-612. Bibl. 4 Agric. Res. Sta. Bhavanisagar Madras, India.

35 ANDERSEN, Alice M.
A preliminary study of dormancy in browntop and cattail millets. Proc. Ass. Seed Anal. N. Amer. 1958:48:85-92.

36 ANDERSEN, A. M.
Effect of gibberellic acid, kinetin-like substance, Ceresan, and phenacridance chloride on the germination of Panicum ramosum seeds. Proc. Int. Seed Test. Ass. 1962, 27: No. 3: 730-741.

37 ANDERSEN, R. N., BEHRENS, R. and LINCK, A. J.
Effects of dalapon on some chemical constituents in sugarbeets and yellow foxtail (Setaria glauca), Weeds 10(1)·4-9 1962.

38 ANDERSEN, R. N., LINCK, A. J. and BEHRENS, R.
Absorption, translocation and fate of dalapon in sugarbeets and yellow foxtail. Weeds 10(1):1-3. 1962.

39 ANDERSON, E.
A semigraphical method for the analysis of complex problems. Proc. Nat. Acad. Sci. Wash. 43:923-927. 1957.

40 ANDERSON, E.
The evolution of domestication In: Evolution after Darwin, Ed. by S. Tex. Univ. Chicago Press, 2, 67. 1960.

41 ANDERSON, J. G.
A new species of Panicum. Bothalia 1967: 9: No. 2: 341-342; from Herb. Absts. 1968 38. Abst. 459.

42 ANDERSON, L. P.
The place for millet and sudan grass. Plant Food 10 (1): 7-8. 1964.

43 ANDREEVA, E. A. and SCHELOVA, G. M.
Uptake of soil and fertilizer nitrogen by plants as revealed by greenhouse pot experiments using N15. (Dockucheav Soil Inst., Moscow, U. S. S. R.). In: Meeting of Commission II and IV of the International Society of Soil Science and Soil Chemistry and Fertility, Sept. 1966, Aberdeen, Scot., Int. Soc. Soil Science; Amsterdam, Neth. 113-124. 1967.

44 ANJANEYULU, V. S. R. and RAO, M. V.
   Mixed cropping as a means of cotton expansion in the Western Tract. Indian Cott. Gr. Rev. 13:459-465. 1959.

45 ANONYMOUS
   Silage from Japanese millet.   Australian Reference Newsletter, March 8, 1948.

46 ANONYMOUS (AFRICA, WEST)
   Les crioceres du mil en Afrique moyens de lutte (Criocerids,
   pests of millet in Africa. Means of Control). Cah. Agric.
   Prat. Pays Chauds 1965 No. 1 [Agron. Trop. 20 (1 suppl.):
   53-54] Paris 1965.

47 ANSLOW, R. C. and BELCOURT, M. S.
   Fodder production from Setaria sphacelata in Mauritius. I.
   Strains and growth habit, establishment, productivity and fer-
   tilizer response. II. Feeding value. Rev. Agric. Sucr.
   Maurice 1958, 37:115-20.

48 APPERT, J.
   Les chenilles mineuses des cereales en Afrique Tropicale
   (Caterpillars mining cereals in Tropical Africa). Agron.
   Trop. 19(1):60-74. Paris, 1964.

49 ARAI, Masao
   Studies on the chemicals effective for breaking seed dorman-
   cy and for killing dormant seed of Echinochloa crusgalli Beauv.
   Var. Oryzicola Ohwi. 1. Screening of effective chemicals.
   Proc. Crop. Sci. Soc. Japan 36(3):321-325. Illus. 1967
   (Jap., E. Sum).

50 ARDASEV, M.
   The cultivation of Setaria italica in the Baskirian A. S. S. R.
   Zemledelie 1957: 4:82-84.

51 ARNOLD, B. L.
   Gahi and Starr millet compared. Miss. Fm. Res. 25: No.
   4, 5. 1962.

52 ARNOLD, B. M.
   Coloration of glumes in Panicum miliaceum. Tr. Byuro
   Prikl. Bot. 7(5):293-305. 1941.

53 ARNOLD, B. M.
   A contribution to the classification of Panicum miliaceum L.
   Tr. Prik. Bot. i. Selek. 14(1):252-270. 1924-25.

54 ARNOLD, B. M.
   Effect of duration and time of vegetation of millet on yields.
   J. Exp. Landw. Sudosten. Eur. Russl. 4(1):155-175. 1927.

55 ARNOLD, B. M.
    Millet. Ogiz. Selkolkhozguz, Moscow 1931: pp. 64.

56 ARNY, A. C. , BRIDGEFOR, R. O. and CRIM, R.
    Supplementary and emergency crops for Minnesota.   Minn.
    Agric. Exp. Sta. Bull. 390: 1946. pp. 23.

57 ARNY, A. C. and McGINNIS, F. W.
    Field crop variety trials in the Coon Creek peat experiment-
    al fields, 1919-1925. Minnesota Agric. Exp. Sta. Bul. 228,
    pp. 5-42. 1926.

58 ARSHAD, A. M. and GOULD, F. W.
    Echinochloa crusgalli (L.) Beauv. in North America. Amer.
    J. Bot. : 54: 659. 1967.

59 ARTEM'EVA, (Mme) N. N.
    Kollektsiya prosa na fone is kusstvennogo zarazheniya pyl'-
    noi golovnei (Collection of proso millet on the basis of arti-
    ficial inoculation with loose smut.) Nauch. Sotrud. Vses.
    Inst. Rastenievod. 1959, pp. 233-238, 1959.

60 ATHWAL, D, S,
    Bajra cultivation. Extension Bulletin, No. 1, Series 1.
    Agricultural Information Service, Ludhiana, Punjab (India).
    1962.

61 ATHWAL, D. S.
    Hybrid bajra-1 marks a new era. Indian Fmg. 1965, 15:
    No. 5: 6-7.

62 ATHWAL, D. S.
    Current plant breeding research with special reference to
    Pennisetum. Indian J. Genet. Pl. Breed. 26A (Symposium
    Number), 1966. pp. 73-85.

63 ATHWAL, D. S.
    Genetic studies for the establishment of linkage groups in
    Pennisetum typhoides - Annual Research Report March 1965
    to February 1966. Punjab Agricultural University, Plant
    Breeding Department, Ludhiana, India, 1966.

64 ATHWAL, D. S.
    Recent advances in the breeding of pearl millet. J. Sci.
    Industr. Res. 25: 501-502. 1966.

65 ATHWAL, D. S. and GILL, G. S.
    Inheritance of bristling and purple pigmentation in Pennisetum
    typhoides (Burm.) Stapf and Hubb. J. Res. (India) Vol. III
    (3): 253-259. 1965.

66 ATHWAL, D. S. , GILL, G. S. and GILL, B. S.
    Genetic stocks for linkage studies in Pennisetum typhoides

7   Athwal

(Burm.) Stapf and Hubb. J. Res. (Punjab) 3(2): 122-131.
1966.

67 ATHWAL, D.S. and GUPTA, V.P.
Genetic evaluation of grain and fodder quality of Pennisetum
- Annual Report July 1965 to June 1966.   Punjab Agricultural University, Plant Breeding Department, Ludhiana, India.
1966.

68 ATHWAL, D.S., and LUTHRA, R.C.
S.530 - the bristled bajra that baffles birds.   Indian Fmg.
1964, 14:14, 40.

69 ATHWAL, D.S. and RACHIE, K.O.
Potentialities and future breeding procedures for the improvement of bajra. Indian J. Genet. Pl. Breed. 1963. 23:155-
157.

70 ATHWAL, D.S. and RACHIE, K.O.
The co-ordinated bajra hybrid No. 1 (HB-1).   National Seeds
Corporation Ltd. Bulletin 1:4. 1965 (India).

71 ATHWAL, D.S. and SINGH, Gian
Variability in Kangni-1. Adaptation and genotypic and pheno-
typic variability in four environments. Indian J. of Genet. -
Pl. Breed. Vol. 26(2): 142-152. July, 1966.

72 AVDULOV, N.P.
Karyo-systematische untersuchung der familie gramineen.
Bull. Appl. Bot. Genet-Pl. Breed. Suppl. 43. 1931.

73 AYKROYD, W.R., GOPALAN, C. and SUBRAMANIAN, S.C.
The nutritive value of Indian foods and the planning of satis-
factory diets (Special Report Series No. 42). Nutrition Research Laboratories, Hyderabad. Indian Council of Medical
Research, New Delhi, 1963.

74 AYYALUSWAMI, P., JAGANATHAN, V. and JAYARAMAN, V.S.
Studies on maize, ragi, and cow dung in chick mash. Indian
Vet. J. 1967. 44: 331-335. (State Livestock Res. Stat.
Ramanathaparam District Livestock Farm Cheitinad.)

75 AYYANGAR, G.N.R. and HARIHARAN, P.V.
Chlorophyll deficiencies in Pennisetum typhoides (Stapf and
Hubb). The pearl millet. Madras Agr. J. 1935. 23: 394-
397.

76 AYYANGAR, G.N.R. and HARIHARAN, P.V.
The tillers of pearl millet - Pennisetum typhoideum (Rich).
The Madras Agr. J. 23:474-477. 1935.

77 AYYANGAR, G.N.R. and IYER, M.A.S.
Mixed cropping, a review. Madras Agr. J. 30:3-14. 1942.

78 AYYANGAR, G. N. R. and VIJIARAGHAVAN, C.
Germination tests on millets. Madras Dep. Agric. Yearbook.
1926.

79 BABENKOV, I.
An experiment on seed growing at a collective farm. Kolkhoz. Proizv. 1953: No. 1: 24-25. 1953.

80 BADOVSKII, R. I.
From the plot on to the field of a collective farm. Nauk.
Pered. Opyt Sol'. Khoz. 1956: No. 2:34-37.

81 BAGYARAJ, J. and RANGASWAMI, G.
Studies on the effect of foliar nutrient sprays on rhizo-
sphere microflora of Eleusine coracana. Mysore J. Agric.
Sci. 1967 1 No. 3. 176-186. (Bibl. 18 Microbiol. Div. Univ.
Agric. Sci. Bangalore, India.)

82 BAINS, K. S., ATHWAL, D. S. and GUPTA, V. P.
Combining ability of pearl millet inbreds in relation to genetic
diversity of male sterile lines. J. Res. (India) 1967: 4: 103
196. (Dept. Pl. Breed; Punjab Agric. Univ. Ludhiana, India.)

83 BAJWA, M. A., BAJWA, Azeez Manzur Ahmad and SHAH,
Siraj-ud-din
Anthesis in pearl millet (Pennisetum typhoideum). Scientist
(Pakistan) 4(1/2): 53-61. 1961.

84 BAKKE, A. L.
Control and eradication of European bindweed. Iowa St. Bul.
61. pp. 937-960. 1964.

85 BAKSHI, J. S., RACHIE, K. O. and SINGH, Amarjit
Development of dwarf strains of pearl millet and an assess-
ment of their yield potential. Curr. Sci. 35(14): 355-356.
July 20, 1966.

86 BALASUBRAMANIAN, S. C. and RAMCHANDRAN, M.
Amino-acid composition of Indian foodstuffs. 3. Threonine
and arginine content of some cereals. Indian J. Med. Res.
1957. 45:623-629.

87 BALL, C. R.
Pearl millet. USDA Farmers Bull. 168. 1903.

88 BALLARI, C. P.
Behavior of populations and selections of Panicum miliaceum
and Setaria italica in the semiarid region (pp. 166-170).
Fourth Meeting on Forage Plants held at Anguil (La Pampa)
Experimental Station, 20-22 July 1954. pp. 230.

89 BARNARD, C.
    Herbage plant register.   Pl. Int. Rev. 1967: 4: No. 1:
    17-23.

90 BARNES, D. K. and BURTON, G. W.
    Tropical environment of Puerto Rico useful for studying day-
    length sensitivity in pearl millet.   Crop Sci. 6:212-213.
    1966.

91 BARNETTE, R. M., CAMP, J. P., WARNER, J. D. and GALL,
    O. E.
    The use of zinc sulphate under corn and other field crops.
    Florida Sta. Bul. 292. pp. 51. 1936.

92 BARRET, Gary W.
    The effects of an acute insecticide stress on a semi-enclosed
    grassland ecosystem.   Ecology 49(6):1019-1634. Illus.
    1968.

93 BARVE, D. B.
    Pollination in Bajri (P. typhoides). Agric. Livestock India
    2:103. 1934.

94 BATES, L. S.
    Protein values of some millet samples from India.   Unpub-
    lished personal correspondence with L. S. Bates, CIMMYT,
    Mexico. 17 May, 1968.

95 BATISTA, A. C.
    Outras especies de Phyllosticta recolhidas em pernambuco
    (Further species of Phyllosticta collected in Pernambuco).
    An. Congr. Soc. Bot. Brazil, 1953, 84-88. 1953.

96 BAVRINA, T. V.
    The effects of prolonged darkness on pigments in neutral,
    long-day and short-day plants. Dokl. Akad. Nauk S. S. S. R.
    1966, 1967: No. 2. 464-467.

97 BAXTER, H. D., OWEN, J. R. and RATCLIFF, L.
    Comparison of two varieties of pearl millet for dairy cattle.
    Tennessee Farm and Home Science Progress Report 33: 4-
    5. 1960.

98 BEATY, E. R., HAYES, D. D. and WORLEY, E. E.
    Value of pelleted fescue, brown-top millet, and whole stalk
    corn as steer feed. Agron. J. 1963, 55:531-532.

99 BEATY, E. R., SMITH, Y. C., McCREERY, R. A., ETHREDGE,
    W. J. and BEASLEY, K.
    Effect of cutting height and frequency on forage production
    of summer annuals. Agron. J. 1965, 57: No. 3: 277-279.

100 BECK, E. W. and SKINNER, J. L.
     Screening of insecticides against the changa and the south-
     ern mole cricket attacking seedling millet. J. Econ. En-
     tomol. 60(6): 1517-1519. 1967.

101 BEGG, J. E.
     Rep. Div. Ld. Res. Reg. Survey. C. S. I. R. O. Aust.
     1963-64.

102 BEGG, J. E.
     High photosynthetic efficiency in a low latitude environ-
     ment. Nature 205(4975): 1025-1026. 1965.

103 BEGG, J. E.
     The growth and development of a crop of bulrush millet
     (Pennisetum typhoides S. and H.). J. Agric. Sci. 65:
     341-9. 1965b.

104 BEGG, J. E., BIERHUIZEN, J. F., LEMON, E. R., MISRA,
     D. K., SLATYER, R. O. and STERN, W. R.
     Diurnal energy and water exchanges in bulrush millet in
     an area of high solar radiation. Agr. Meteorol. 1: 294-
     019. 1001

105 BEJDENKOVA, A. F. and KONOVALOV, I. N.
     An investigation of the physiological effect of growth sub-
     stances on plants. Proc. 9th Internat. Bot. Congr. 1959,
     2:27.

106 BEL'DENKOVA, A. F.
     The influence of gibberellic acid on the growth and devel-
     opment and morphological variability in plants. Tr. Bot.
     Inst. V. L. Komarova (4) 1962, 15: 101-119.

107 BELIKOV, M. N. and CHUBOV, P. O.
     Mass-intoxication of fine-wooled sheep and lambs by ordi-
     nary millet. Sbornik Nauc. Issled. Rabot Stud. Stavropol'-
     sk. S-Kh. Inst. 1956 (4): 124-125. 1956.

108 BELLIS, E.
     Annual Report for Irrigation Investigations, 1957. Rep.
     Dep. Agric. Kenya Vol. 2:110-115. 1959.

109 BELOV, S. A.
     A study of the genus Panicum. Tr. Byuro Prikl. Bot.
     7(5): 306-324. 1914.

110 BELOV, S. A.
     Contribution to the study of Panicum miliaceum. Bezen-
     chuk. Selsk. Khoz. Opyin. Sta., No. 73:333-352. 1916.

111 BELOVA, S. M. and DENISENKO, Ya I.
     The chemical nature of prosol. Prikl. Biokhim. Mikro-

biol. 1(6): 664-668. 1965.

112 BELOVA, S. M. and DENISENKO, Ya I.
     Spectrophotometric determination of linoleic and linolenic
     acids in millet oil.   Prikl. Biokhim. Mikrobiol. 1(4):
     474-476. 1965.

113 BELOVA, S. M. and DENISENKO, Ya I.
     Vitamin composition of millet oil.   Prikl. Biokhim. Mikro-
     biol. 1(4): 387-390. 1965.

114 BENGTSSON, A. and others
     Silage plants for sandy soils.   Medd. 93 Stat. JordbrFör-
     sök 1958, pp. 125.

115 BENNETT, R. L. and IRBY, G. B.
     Late crops for overflow lands.   Arkansas Sta. Bull. 27,
     pp. 53-61.   Exp. Sta. Rec. 6:212.   1894-1895.

116 BERKNER
     The millet and sorghum species and their cultivation.
     Dtsch. Landw. Pr. 1940.   67:p. 29.

117 BERTHOLON, P.
     The millets and their grinding in French West Africa.
     Bull. Anciens Eleves Ecole Francaise Meunerie 109:17-22.
     1949.

118 BEYERS, C. P. del
     The influence of copper fertilization on the copper content
     of certain crops.   S. Afr. J. Agric. Sci. 1966: 9: No.
     4.   907-910.
     (Bibl. 5: Af. e. Fv. Landb. NAVOR Inst.)

119 BEZOT, P.
     L'amelioration des cultures cerealieres au chad.   Agron.
     Trop. 18(1): 128-131.   Jan. 1963.

120 BHALLA, S. K. and ATHWAL, D. S.
     A study of heterosis in some inter-varietal and top cross-
     es of pearl millet. J. Res. (India) 1968: 5: No. 2: 58-62.

121 BHATAWADEKAR, P. U. , CHINEY, S. S. and DESHMUKH,
     K. M.
     Response of Bajri-turmixed crop to nitrogen and phosphate
     fertilization under dry farming conditions of Sholapur.
     Indian J. Agron. 11:243-246.   1966.

122 BHATIA, D. R.
     Observations on the biology of the desert locust (Schisto-
     cerca gregaria Forsk.) in Sind-Rajputana Desert area.
     I. The preferred food plants of the locust.   Indian J. Ent.
     2(2): 187-192.   N. Delhi, 1940.

123 BHATT, B. Y. et al.
Some aspects of irradiation of seeds with ionizing radiations. (pp. 591-607) (from "Effects of Ionizing Radiations on Seeds). International Atomic Energy Agency, Vienna (1961): Sch. 199. 50: pp. 655.

124 BHATT, R. S.
Proceedings of conference of workers on millets, Kolhapur. 1955. Indian Council of Agricultural Research, p. 146. 1955.

125 BILQUEZ, A. F., MAGNE, C. and MARTIN, J. P.
Bilan de six annees de recherches sur l'amalioration des plantes au Senegal. (pp. 585-601) Dept. of the FAO-IAEA Technical Meeting, Rome, 22 May - 1 June, 1964. Pergamon Press, London, pp. 832, 1965.

126 BINDRA, O. S. and KITTUR, S. U.
Mermis sp. (Nematoda) parasitising caterpillars of Amsacta moorei Butler. Indian J. Ent. 18 (4): 462-463. 1956.

127 BIRCH, W. R. and BOMMEGOWDA, A. M.
The effect of soil drying on millet.   Pl. Soil 24(2):333-335. 1966.

128 BIRCHFIELD, Wray
Histopathology of nematode-induced galls of Echinochloa colonum.   Phytopathology 54(8):888.   1964.

129 BISSELL, T. L.
Army worms in Georgia.   J. Econ. Entomol.   37(1): 112-113. 1944.

130 BLAIR, R. E.
Report of field crops work on the Yuma Reclamation Project Experiment Farm in 1916.   USDA Bur. Pl. Indus., Work Yuma Exp. Farm, 1916, pp. 13-19, 22-31. 1916.

131 BLATTER E. and McCANN, C.
The Bombay grasses.   Sci. Monogr. Indian Coun. Agric. Res. 5: 177-184.   1935.

132 BODA, J.
Induction of tobacco, millet and mustard mutations by means of X-irradiation.   Novenytermeles 13: 319-338. 1964.

133 BOGDAN, A. V.
Hybridization in the Setaria sphacelata complex in Kenya. (pp. 311-317).

134 BOIKO, L. A. and MATUKHIN, G. R.
     Enzymatic reactions of some carbohydrates in the leaves
     of cultivated grasses which occur with their adaptation to
     salinized soil.   Nauch. Dokl. Vyss. Shk. Biol. Nauk. 2:
     154-157. 1964.

135 BONO, M.
     Varietal experimentation with millet (Pennisetum).   The
     problems, the results and a new orientation of breeding
     work. Ann. Centre Rech. Agron. Bambey 1960-61: No.
     20:119-36.

136 BONO, M.
     Contribution a L'etude Botanique Du Mil Pennisetum.
     CCTA/FAO Symposium on Savannah Zone Cereals.   Dakar.
     29 August - 4 September, 1962.

137 BONO, M. and others
     Contribution a l'etude des populations selectionees de mil
     (Pennisetum typhoides). Bull. Agron. Minist. Fr. d'Out.
     Mer 1957, No. 15:49-57, Bibl. 5 (Cent. Rech. Agron.,
     Bambey, Senegal).

138 BOR, N. L.
     Two confused species of Panicum.   Feddes Rep.   1960.
     63:328-331.

139 BOR, N. L.
     The grasses of Burma, Ceylon, India and Pakistan (ex-
     cluding Bambuseae).   Pergamon Press.   London.   1960.
     pp. 262-265: 309-311: 340-343: 350-351: 362-363.

140 BORCHHARDT, A.
     Operations of the phytopathological section of the agricul-
     tural experiment station in the eastern steppe region (Eka-
     terinoslaff Govt.) in the year 1925.   Dnepropetrovsh.
     pp. 3-37 (cf. Rev. Appl. Mycol. 7:223) 1927.

141 BOUYER, S.
     Contribution a l'etude agrologique des sols du Senegal
     (Casamance exceptee).   Contribution to the agronomic
     study of the soils of Senegal (Casamance excluded) Bull.
     Agric. Congo Belge 40: 887-1020.

142 BOWDEN, J.
     New species of African stemboring agrostidae (Lepidop-
     tera). Bull. Ent. Res. 47(3):415-428. London, 1956.

143 BOYLE, J.W. and JOHNSON, R. I.
     Pearl millet--new summer forage crop in N. S. Wales.
     Agric. Gaz. N. S. W. 1968, 79, No. 9:513-515.
     (Agron Sect. , Dept. Agric. , Sydney, Australia).

144 BRACK, A., BRUNNER, R. and KOBEL, H.
The course of alkaloid synthesis in saprophytic cultures
by a strain of Claviceps from Pennisetum typhoideum.
Arch. Pharm. u. Ber. Deutsch. Pharmazeut. Gesell.
295(7): 510-515. 1962.

145 BRANDON, J. F. et al.
Prose or hog millet in Colorado. Colo. Agric. Exp. Sta.
Bull. 383. 1932.

146 BREAKWELL, E.
The grasses and fodder plants of New South Wales. Sydney, A. J. Kent, Government Printer, 1923.

147 BRIGGS, L. J. and SHANTZ, H. L.
The water requirement of plants. USDA Bur. Pl. Indus.
Bull. 284, pp. 49: 285. pp. 96. 1910-1911.

148 BRIGGS, L. J. and SHANTZ, H. L.
Relative water requirement of plants. J. Agr. Res. 3(1):
1-64. 1914.

149 BRIGGS, L. J. and SHANTZ, H. L.
Daily transpiration during the normal growth period and
its correlation with the weather. J. Agr. Res. 7(4): 155-
212. 1916.

150 BRINDLEY-RICHARDS, G. J.
Clonal studies on Setaria sphacelata. Agric. Res. (Pretoria) 1963, 483-484.

151 BROD, G.
Studies on the biology and ecology of barnyard grass,
Echinochloa crus.-galli (L) Beauv.
(Pflanzent zaut, Karlsrue, West Ger.)
Weed Res. 8(2): 115-127. Illus. 1968.

152 BROMLEY, M. and BARRAU, J.
Existence of a cultivated Coix in the mountains of New
Guinea. J. Agric. Trop. Bot. Appl. 1965, 12:781-782.

153 BROOK, O. L. and BEATY, E. R.
Yield and per cent leaves, stems and grain of crops grown
for silage in S. E. Georgia. Ga. Agric. Res. 1967, 8,
No. 4, 3-5 (Bibl. H. Georgia Agric. Exp. Sta. Midville).

154 BROOKHAVEN NATIONAL LABORATORY
Annual report 1 July, 1967 pp XXVI and 227

155 BROWN, K. J.
Rainfall, tie-ridging and crop yields in Sukumaland,
Tanganyika. Emp. Cott. Gr. Rev. 1963, 40: No. 1:34-
40.

156 BRUNNICH, J.C.
     Hydrocyanic acid content of millets and sorghums.
     Queensland Dept. Agr. and Stock, Ann. Rpt., 1923-24:
     pp. 31-33.   1924.

157 BRUYÈRE, R.
     Cereales d'altitude.   Bull. Ineac 1958, 7:53-55.

158 BRUYN, J.A. de
     The in vitro germination of pollen of Setaria sphacelata.
     1. Effects of carbohydrates, hormones, vitamins and
     micronutrients. Physiol. Plant. 1966. 19:322-327.

159 BRUYN, J.A. de
     The in vitro germination of pollen of Setaria sphacelata.
     2. Relationships between boron and certain cations.
     Physiol. Plant. 1966. 19: 322-327.

160 BUCK, J.L.
     Land utilization in China.   The Council of Economic and
     Cultural Affairs, Ind. New York, 1956.

161 BUCK, J.L., DAWSON, O.L. and WU, Yu L.
     Food and Agriculture in Communist China (The Hoover In-
     stitution on War, Revolution and Peace - Stanford Univer-
     sity, Palo Alto, Calif.) Frederick A. Praeger, Publish-
     ers. New York. 1966.

162 BUCKLEY, T.A. and ALLEN, B.F.
     Notes on current investigations, April to June 1951.
     Malay. Agric. Jour. 34: 133-141. 1951.

163 BURGER, A.W. and HITTLE, C.V.
     Yield protein, nitrate and prussic acid content of sudan-
     grass, sudangrass hybrids and pearl millets harvested at
     two cutting frequencies and two stubble heights.   Agron.
     J. 1967, 59. No. 3, 259-262 (Bibl. B; Dept. Agron. Univ.
     Illinois, Urbana).

164 BURGESS, A.P.
     Calories and proteins available from local sources for
     Uganda Africans in 1958 and 1959.   E. African Mod. J.
     1962, 39:449-463.

165 BURKILL, I.H.
     A dictionary of economic products of the Malay Peninsula.
     Crown Agents for the Colonies, London, 1935.

166 BURNS, D.B.
     Rep. Dep. Agric. Bombay. 1922-1923: 156. 1924.

167 BURNS, D.B.
     Rep. Dep. Agric. Bombay. 1923-1924: 147. 1925.

168 BURNS, W.
Technological possibilities of agricultural department in India. Supt. Gov't. Printing, Lahore. 1944.

169 BURTON, G.W.
A cytological study of some species of the tribe Paniceae. Amer. J. Bot. 29 (5): 355-359. 1942.

170 BURTON, G.W.
Immediate effect of gametic relationship upon seed production in pearl millet, Pennisetum glaucum. Agron. J. 44:424-427. 1952.

171 BURTON, G.W.
Gahi-1 Pearl Millet - tops for temporary summer pasture. The Progressive Farmer (Georgia-Alabama-Florida edition). May, 1960.

172 BURTON, G.W.
1964 Annual Report - Grass Breeding Project, Tifton, Georgia (unpublished).

173 BURTON, G.W.
The current status of pearl millet breeding. Published in the Report of 21st Southern Pasture and Forage Crops Improvement Conference, Gainesville, Florida, April 22-23, 1964.

174 BURTON, G.W.
Male-sterile pearl millet Tift 18A released. Crops Soils 1965. 18: No. 1: p. 19.

175 BURTON, G.W.
Photoperiodism in pearl millet, Pennisetum typhoides. Crop Sci. 1965. 5: 333-335.

176 BURTON, G.W.
Pearl millet Tift 23A released. Crops Soils 1965, 17: No. 7: p. 19.

177 BURTON, G.W.
Pearl millet breeding. Sols Afr. 11(1 and 2): 39-42. 1966.

178 BURTON, G.W.
Prospect for the future. Plant Breeding - A symposium held at Iowa State University. University Press, Ames, Iowa. 1966.

179 BURTON, G.W.
Genetic variances in pearl millet (Pennisetum glaucum) and their significance to the forage breeder. The X International Congress of Genetics, Proceedings, Vol. II.

180   BURTON, G.W.
      Quantitative Inheritance in Grasses. Reprinted from the
      Proceedings of the Sixth International Grassland Congress.

181   BURTON, G.W.
      For forage and food: Dwarf pearl millet.   Agric. Res.
      (Wash.) 16(1):4.   July, 1967.

182   BURTON, G.W.
      Heterosis and heterozygosis in pearl millet forage produc-
      tion. Crop Sci. 8(2): 229-230. Mar/Apr. 1968.

183   BURTON, G.W. and ATHWAL, D.S.
      Two additional sources of cytoplasmic male sterility in
      pearl millet and their relationship to Tift 23-A.   Crop
      Sci. 7: 209-211.   1967.

184   BURTON, G.W. and FORTSON, J.C.
      Inheritance and utilization of dwarfism in pearl millet
      (Pennisetum glaucum). (pp. 64-65).   Proc. Ass. South.
      Agric. Wkrs., 61st Ann. Conv., Atlanta, Ga., Feb. 3-5,
      1964, pp. 299.

185   BURTON, G.W. and FORTSON, J.C.
      Lattice-square designs increase precision of pearl millet
      forage yield trials. Crop Sci. 1956. 5: p. 595.

186   BURTON, G.W. and FORTSON, J.C.
      Inheritance and utilization of five dwarfs in pearl millet
      (Pennisetum typhoides) breeding. Crop Sci. 1966, 6:69-
      72.

187   BURTON, G.W., GUNNELS, J.B. and LOWREY, R.S.
      Effect of maturity on the yield and quality of pearl millet.
      (p. 9).   Abstracts of the Annual Meetings of the American
      Society of Agronomy, Columbus, Ohio, 31 Oct. to 5 No-
      vember 1965, pp. 134.

188   BURTON, G.W. and HART, R.H.
      1962 Annual Report: Grass Breeding Project - Tifton,
      Georgia.

189   BURTON, G.W. and JACKSON, J.E.
      Managing pearl millet for meat and milk production.   Ga.
      Exp. Sta. Leaflet 25. 1961.

190   BURTON, G.W., KNOX, F.E. and BEARDSLEY, D.W.
      Effect of age on the chemical composition and digestibility
      of grass leaves. Agron. J. 1964, 56: No. 2:160-161.

191   BURTON, G.W. and POWELL, J.B.
      1963 Annual Report: Grass Breeding Project - Tifton,
      Georgia.

192 BURTON, G. W. and POWELL, J. B.
     Six chlorophyll-deficient seedlings in pearl millet, Penni-
     setum typhoides, and a suggested system for their nomen-
     clature. Crop Sci. 5:1-3. 1965.

193 BURTON, G. W. and POWELL, J. B.
     Genetic and cytogenetic analysis of the effects of recur-
     rent irradiation and chemical mutagens on general and spe-
     cific combining ability in pearl millet, Pennisetum ty-
     phoides (P. glaucum). Unpublished Progress Report No.
     4 for 1 May, 1965, to 30 April, 1966, at Tifton, Georgia.

194 BURTON, G. W. and POWELL, J. B.
     Morphological and cytological response of pearl millet,
     Pennisetum typhoides, to thermal neutrons and ethyl me-
     thane sulfonate seed treatments. Crop Sci. 6:180-182.
     1966.

195 BURTON, G. W.
     Pearl millets Tift 23DA and 23DB released. Ga. Agric.
     Res. 1967. 9. No. 1. p. 6.

196 BURTON, G. W. , GUNNELLS, J. B. and LOWRY, R. S.
     Yield and quality of early and late maturing, near isogenic
     populations of pearl millet. Crop Sci. 8(4): 431-434.
     1968.

197 BURTON, G. W.
     Epistasis in pearl millet forage yields. Crop Sci. 8(3):
     365-368. 1968.

198 BURTON, G. W. and ATHWAL, D. S.
     Reciprocal maintainer-restorer relationship between A_1 and
     A_2 sterile cytoplasm facilitates millet breeding. Univ. Ga.
     Coll. Agric. Exp. Sta. ; Tifton, Georgia.

199 BURTON, G. W. and POWELL, J. B.
     Genetic and cytogenetic analysis of the effects of recur-
     rent irradiation and chemical mutagens on general and spe-
     cific combining ability in pearl millet (Pennisetum ty-
     phoides) (P. glaucum). Progress Report 1 May, 1966 -
     30 April, 1967. Contract AT (40-1) 2976: pp 83 from
     Nuclear Sci. Abst. 1967: 21: Abst. 16312.

200 BURTON, G. W. and POWELL, J. B.
     Genetic and cytogenetic analysis of the effects of recur-
     rent irradiation and chemical mutagens on general and spe-
     cific combining ability in pearl millet Pennisetum ty-
     phoides. Three year review 1 May, 1965 - 30 April,
     1968. Contract AT (40-1) 2976: pp. 2976: pp. 17 from
     Nuclear Sci. Abst. 1968: 22: Abst. 23488.

201 BUSSE, W.
The influence of naphthalin on the germination of cereals. Abs. in Zischr. Pflanzenkrank., 14 (1904), No. 4, pp. 219-220.

202 BUSSON, F.
Plantes alimentaires de l'Ouest Africaine. Etude Botanique, Biologique et Chimique. Editions Leconte, Marseilles, France 1965, 120 F.

203 BUTLER, E. J.
Some diseases of cereals caused by Sclerospora graminicola, Mem. Dep. Agric. India Bot. Ser. 2:1-24. 1907.

204 BUTLER, E. J.
Fungi and disease in plants. PP. VI + 547. Thacker Spink and Co. Calcutta. 1918.

205 BUTLER, E. J. and BISBY, G. R.
Fungi of India, Sci. Monogr. Indian Coun. Agric. Res. 1931.

206 CAIRNIE, A. G. and MONSIGLIO, J. C.
Chemical composition of native or introduced forage species in the semi-arid region of La Pampa. Revta Invest. Agropec. Ser 2. 1967, 4, No. 11, 207-21. (Bibl. 32; Es. e Estac. Exp. Agropec. Anguil, La Pampa, Argentina).

207 CALDUCH, M.
Plants of my herbarium--Note on the genus Setaria. Collnea Bot. Barcinone 1968, 7. No. 1, 151-63 (Spanish) (Morph).

208 CALLEN, E. O.
Food habits of some pre-Columbian Mexican Indians. Econ. Bot. 1965, 19:335-343.

209 CAMUS, A.
New Panicum species from Madagascar. Bull. Soc. Bot. Fr. 1962, 99:63-65.

210 CANOV, T. and KODANEV, I.
The Gorjkii State Agricultural Research Station. Kolkhoz. Proizv. No. 1:57-59. 1953.

211 CARBIENER, R., JAEGER, P. and BUSSON, F.
Etude de la fraction protidique de la graine de fonio Digitaria exilis (Kippist) Stapf, une proteine exceptionnellement riche in methionine. Ann. Nutrition Alimentation, 1960, 14: Mem 165-169.

212 CARDER, A.C.
Growth and development of some field crops as influenced by climatic phenomenon at two diverse latitudes. Can. J. Pl. Sci. 37(4): 392-406. 1957.

213 CARNE, W.M.
Smut on broom millet and other sorghums. J. Dep. Agr. West. Aust. 2(4): Ser. 3 pp. 348. 1927.

214 CARTER, J.
Field experiments at Las Vegas, 1937-1939. New Mexico St. Bul. 271, pp. 11. 1940.

215 CASTELLANI, Ettore
Anthracnose of Eragrostis teff. Nuovo Gior. Bot. Ital. 55(1): 142-144. 1948.

216 CASTELLANI, E. and CICCARONE, A.
Malattie crittogamiche del 'Teff' (Fungal diseases of 'Tef') Florence, Regio Instituto Agronomico per L'Africa Italiana, 1939.

217 CASTELLANI, E.
Un accoumie togliare del Coix lachryma-jobi L. (Leaf blotch of Coix lachryma-jobi L.) Riv. Agric. Subtrop. Trop. 1952, 46, No. 4-6. 159-65, Bibl. 19.

218 CATHERINET, M.D., GABORIT, J. and MAYAKI, A.A.
The question of millet breeding in Niger. Agron. Trop., Paris 1963. 18:119-125.

219 CENTRAL ARID RESEARCH INSTITUTE
Response of Pennisetum typhoideum to minor nutrients. Agric. Res. (India) 2(1): 60. 1962.

220 CEYLON DIRECTOR OF AGRICULTURE
Administration Report for 1960. Part IV. Education, Science and Art (C). Ceylon Adm. Rep. Agric. 1960 (1962): pp. 278.

221 CHAKRAVARTI, B.P. and GHUFRAN, S.M.
Ozonium wilt of marua in Bihar. Proc. Bihar Acad. Agric. Sci. 12-13(1): 51-52. 1964.

222 CHALUPA, W., BRANNON, C.C. and CONLEY, C.
Hybrid sorghum and pearl millet as supplementary grazing crops for dairy cattle. S. Carolina Agric. Exp. Sta. Bull. No. 526, July 1966 p. 14.

223 CHAND, J.N. and SINGH, B.
A new Helminthosporium disease of bajra. Curr. Sci. 39(9): 240. 1966.

21   Chandrasekhara

224 CHANDRASEKHARA, M.R., RAO, M.N., SWAMINATHAN, M.,
    LAHIRY, N.L., BHATIA, D.S., PARPIA, H.A.B. and SUB-
    RAMANYAN, V.
    Production of malt extract--project costs.   Res. and Indry.,
    1959. Vol. 4(8): 789-191. 1959.

225 CHANG, L.P.
    Studies on flowering and hybridization technique in Setaria.
    Acta Agr. Sin. 1958, 9:68-76.

226 CHANNAMA, K.A. and DELVI, M.
    Effects of seed treatment on the viability of jowar (Sor-
    ghum vulgare Pers) and ragi (Eleusine coracana Gaertn.)
    seeds in storage. Inst., Hebbal, Bangalore, Mysore, In-
    dia) Bull. Indian Phytopath. Soc. 3:46-49. 1966.

227 CHARREAU, C. and VIDAL, P.
    Influences of Acacia albinda on soil, mineral nutrition and
    yield of Pennisetum millet in Senegal. Agron. Trop.
    Paris 1965, 20: No. 6-7: 600-626.

228 CHASE, A.
    The Linnaean concept of pearl millet. Amer. J. Bot.
    8(1): 41-49. 1921.

229 CHATTERJEE, B.N. and RICHHARIA, R.H.
    Studies on some of the promising strains of Pennisetum
    pedicellatum. Trin. Proc. Bihar Acad. Agric. Sci. 1955:
    4: 126-128.

230 CHATURVEDI, S.N. and KOUL, A.K.
    Some unreported hosts of Helminthosporium. Agra. Univ.
    J. Res. Sci. 12(1): 59-61. 1963.

231 CHAVAN, V.M.
    Rep. Dep. Agric. Bombay 1936-1937: 200. 1936-37.

232 CHAVAN, V.M.
    Bajra cultivation in India. Indian Council of Agricultural
    Research Farm Bulletin No. 48 (1958).

233 CHEN, C.C. and HSU, C.C.
    Cytological studies on Taiwan grasses. I. Tribe Pani-
    ceae. Bot. Bull. Acad. Sin. 1961, 2:101-10.

234 CHERIAN, Jacob K.
    A new millet - Brachiaria ramosa Stapf. Madras Agr. J.
    31:12-14. 1943.

235 CHERNYAVSKAYA, Z.S.
    Results of work in selection and seed production with mil-
    let and Polygonum. Tr. Ukrain. Nauch. -Issledovatel.
    Inst. Rastenievodstva, Selektsii i Genet. 3:83-93. 1959.

236 CHESALIN, G. A. and SHEHEGLOV, Yu V.
The development of agents for the chemical control of weeds in sowings of sudangrass, foxtail millet and ryegrass for seed production. Tr. Vses. Nauchn. -Issled. Inst. Udobrenii i Agropochvovedeniya 39:79-92. 1962.

237 CHEVALIER, A.
A disease of pearl millet in Senegal. Rev. Bot. Appl. 11:40-50. 1931.

238 CHEVALIER, A.
Sur l'origin des Digitaria cultives. Rev. Bot. Appl. 1950, 30: 329-330.

239 CHEVAUGEON, J.
Maladies des plants cultivier en moyennes-Casamance et dans la delta central nigarien. Rev. Path. Veg. 31:3-51. 1952.

240 CHHONKAR, R. P. S., SACHDEVA, K. K., SINGH, G. S. and SINGH, S. N.
Studies on the varietal differences in the chemical composition of jowar (Sorghum vulgare) and bairn (Pennisetum typhoides) grown in the region of Uttar Pradesh. Balwant Vidyapeeth J. Agric. Sci. Res. 1964 (1967) 6. No. 1 and 2:14-25.

241 CHILCOTT, E. C. and SAUNDERS, D. A.
Millet. S. Dak. Sta. Bul. 60, pp. 127-140. Exp. Sta. Rec. 10:629. 1898-1899.

242 CHINA AGRICULTURE
The production of Panicum millet 142 and its extension. Hsi-pei Nung-yeh Ko-hsueh 1958, No. 3:165-67.

243 CHISAKA, Hideo, TAKAYOSHI, Kataoka and ARAI, Masao
Studies on chemicals effective for breaking seed-dormancy, and for killing dormant seed of barnyard grass (Echinochloa crus-galli.
Beauv. var. oryzicola Ohwi)
(2) Factors influencing the activity of effective chemicals in the soaking treatment. Proc. Crop. Sci. Soc. Japan 36(3): 327-333. Illus. 1967 (J. Eng. sum).

244 CHISAKA, Hideo, TAKAYOSHI, Kataoka and ARAI, Masao
Studies on chemicals effective for breaking seed-dormancy and for killing dormant seed of barnyard grass (Echinochloa crus-galli.
Beauv. var oryzicola Ohwi).
(3) Activity of several chemicals used in contact with soil. Proc. Crop Sci. Japan 36(3): 332-337. Illus. 1967. Jap. Eng. Sum.

245 CHITRE, R.G., DESAI, D.B. and RAUT, V.S.
The nutritive value of pure bred strains of cereals and
pulses. Part I. Thiamine, riboflavin and nicotinic acid
contents of 107 pure bred strains of cereals and pulses.
Indian J. Med. Res. 43:4:575-583. 1955.

246 CHITTENDEN, F.H.
Insect injury to millet. Proc. 10th Mtg. Ass. Econ. Ent.
(USDA Div. Ent. Bul. 17) pp. 84-86. Exp. Sta. Rec. 10:
1061. 1898-1899.

247 CHOWDHURY, S.
Some studies on the smut Ustilago coices Br. of Job's
tears millet. J. Indian Bot. Soc. 25:123-130. 1946.

248 CHOWDHURY, S.
A disease of Zea mays caused by Colletotrichum gramini-
colum. Indian J. Agric. Sci. 6:833-843. 1936.

249 CIFERRI, R. and BALDRATI, I.
(1) Cereali dell'Africa Italiana. (2) Il "Teff" Eragrostis
tef cereale da panificazione dell'Africa orientale Italiana
montana. Firenze: Instituto Agronomico per L'Africa
Italiana 1939.

250 CIRK, G.P.
Extending the range of millet towards the north. Soviet
Plant Industry Record 1940. No. 2:108-111.

251 CLARK, N.A., HEMKEN, R.W. and VANDERSALL, J.H.
A comparison of pearl millet, sudangrass and sorghum-
sudangrasshybrid as pasture for lactating dairy cows.
Agron. J. 57(3):266-269. 1965.

252 CLIFFORD, H.T.
Vivipary in Eleusine indica (L.) Gaertn. (u. Queensland,
Brisbane). Nature 184 (Suppl. 24): 1888. 1959.

253 CLINTON, G.P.
Grain smut of Hungarian grass. pp. 347, 348. Illinois
Sta. Bul. 57. pp. 289-360. Exp. Sta. Rec. 12:357 and
355. 1900-1901.

254 COCHEME, J. and FRANQUIN, P.
FAO/UNESCO/WMO Interagency Project on Agroclimatol-
ogy. Technical Report on a study of the agroclimatology
of the semi-arid area South of the Sahara in West Africa.
Rome: FAO, Paris: UNESCO, Geneva: WMO. p. 332
(Bibl. at ends of sections. Eng. with Fr.).

255 COETZEE, Victoria and BOTHA, Helen J.
A redescription of Hypsoperine acronea (Coetzee, 1956)
sledge golden, 1964 (Nematoda: Heteroderidae), with a

note on its biology and host specificity.   Nematologica 11
(4): 480-484.   1965.

256 COLEMAN, L.C.
        The cultivation of ragi in Mysore.   Bull. Dep. Agric.
        Mysore Entomol. Ser. 11.   1920.

257 COLEMAN, L.C. et al.
        Field crops experiments in Mysore.   Mysore Dep. Agr.
        Rpt. 1921-1922 pts. 1, pp. 4-7, 8 and 9.   Pt. 2 pp. 6,
        7, 15-24, 26-32, 55-59, 60-62.   1922.

258 COMMONWEALTH (Brisbane)
        Rept. Div. Trop. Past., C.S.I.R.O. Aust. 1961.   62:
        pp. 35: 1962-1963: pp. 48.

259 COMPTON, R.H.
        A further contribution to the study of right and left-hand-
        edness.   J. Genetics 2(1):53-70.   1912.

260 CONGO-L'INEAC
        Report for the years 1944 and 1945.   Publ. Inst. Nat.
        Agron. Congo Belge 1947: pp. 101.

261 CONGO-L'INEAC
        Annual Report for the financial year 1950.   Publ. Inst.
        Nat. Agron. Congo Belge 1951: Hors Ser. pp. 392.

262 CONGO-L'INEAC
        Rapport Annuel pour L'Exercice 1952.   L'INEAC Gemb-
        loux, 1953, pp. 395 (Belgian Congo).

263 CONGO-L'INEAC
        Rapport Annuel Pour L'Exercise 1953.   (Annual Report for
        1953).   Institut National pour L'Etude Agronomique du
        Congo Belge.   Bruxelles, 1954.

264 CONGO-L'INEAC
        Rapport Annual Pour Exercise 1954.   (Annual Report,
        1954).   Institut National pour L'Etude Agronomique du
        Congo Belge.   Bruxelles, 1955.

265 CONGO-L'INEAC
        Annual Report for 1955.   Brussels, pp. 567, 1956 (G.
        Eleusine sp. at Kiyaka p. 413).

266 CONGO-L'INEAC
        Rapport Annuel pour L'Exercice 1956.   (Annual Report for
        1956).   Institut National pour L'Etude Agronomique du Con-
        go Belge.   Bruxelles, 1958.

267 CONGO-L'INEAC
        Annual Report for 1956.   Brussels, pp. 548, 1958.   (4.

Eleusine at Kiyaka, p. 386).

268 CONGO-L'INEAC
Du Congo Belge. Eleusine coracana. Rep. L'Ineac 1958-
1959, 451, 468 and 497.

269 COOKE, T.
The flora of the presidency of Bombay 2:907-17. Taylor
and Francis, London. 1908.

270 COOPER, R.B. and BURTON, G.W.
Effect of pollen storage and hour of pollination on seed
set in pearl millet, Pennisetum typhoides, Crop Sci. 1965.
5:18-20.

271 CORBETTA, G.
The smut of Panicum crusgalli and P. erectum produced
by Sorosporium bullatum Schroet. Phytopath. Z. 22:275-
280 (cf. Rev. Appl. Mycol. 34: 280).

272 COUNCIL OF SCIENTIFIC and INDUSTRIAL RESEARCH -
NEW DELHI
The Wealth of India: a dictionary of Indian raw materials
and industrial products. Vol. 7: N-Pe, New Delhi:
Council Sci. Indust. Res. 1966 p. 330 (Bibl. in text).

273 CRAIGMILES, J.P. and ELROD, J.M.
Browntop millet in Georgia. Ga. Exp. Sta. Leaflet 14.
1957.

274 CRAIGMILES, J.P., ELROD, J.M. and LUTTRELL, E.S.
The influence of method of seeding and seed treatment on
stands and yields of sudangrass and millet. Ga. Agric.
Exp. Sta. Mimeo Ser. N.S. 47, 1948.

275 CRAWFORD, J.M.
The industries of Russia: Agriculture and Forestry. Vol.
III. Dept. of Agric. of Russia Ministry of Crown Do-
mains: For the World's Columbian Exposition at Chicago.
St. Petersburg, 1893.

276 CROWDER, L.V. and MICHELIN, A.
La produccion de las variedades de pasto Sudan, sorgos y
millo "perla" en el valle del cauca (the yield of varieties
of sudan grass, sorghum and pearl millet in the Cauca
Valley (Colombia). Agric. Trop. 1958, 14(12): 1744-1749.

277 CROWELL, Ivan H.
(Macdonald Coll. Quebec). Two new Canadian smuts.
Can. J. Res. Sect. C. Bot. Sci. 20(5): 327-328. 6 fig.
1942.

278 CSAK, Z.
Mohar (Setaria italica L.) pp. 273-279. The Results of
the National Varietal Trials of Improved Crop Varieties,
1962. Budapest - (Orszagos Mezogazdasagi Fajta-Es Ter-
melestechnikai Minosito Intezet). pp. 372.

279 CUMMING, G.B.
Revisionary studies in the tropical American rusts of Panicum, Paspalum and Setaria. Mycologia 34:683-686. 1942.

280 CUNDEROVA, A.I. and SULYNDIN, A.F.
Millet sickness and its cause. Zemledelie (Agriculture)
1964, No. 2:84-88.

281 CUNNINGHAM, G.H.
Additions to smut fungi of New Zealand. I. Trans. Roy.
Soc. N.Z. 75: 336-339. 1945.

282 CURTIS, D.L., BURTON, G.W. and WEBSTER, O.J.
Carotenoids in pearl millet. Crop Sci. 6:300-301. 1966.

283 CYBUL'KO, V.S.
Twenty-four hour variations of assimilation products in the
leaves of long- and short-day plants. Fiziol. Rast. 1962.
9: No. 5: 567-574.

284 DALELA, G.G.
Experiments on physiologic specialization in Puccinia penniseti Zimm. Indian Phytopathol. 17(1): 63-65. 1964.

285 DANCIK, J.
Growing of millets and legumes: Grass under seed after
winter crops in the maize producing region. Ved. Pr.
Vyzk. Ustavu Rostlinne Vyroby Piestanoch 4:97-111. Illus.
1966 (E. and R. Summ.).

286 DANGE, S.R.S. and KOTHARI, K.L.
Effect of relative humidity and length of storage period upon fungal invasion and germination percentage of bajra
seeds. Bull. Grain Technol. 6, 2, p. 110-112 (1968).

287 DANIEL, V.A., DESAI, B.L.M., SUBRAMANYA, T.S., RA-
JURS, S., RAO, Vanket, SWAMINATHAN, M. and PARPIA,
H.B.A.
The supplementary value of Bengal gram, red gram, soya
bean as compared with skim milk powder to poor Indian
diets based on ragi, Kaffir corn and pearl millet. J.
Nutr. Diet. 5(4): 283-291. 1968.

288 DANIEL, V.A., URS, T.S.R., DESAI, B.L.M., RAO, S.V.,
and SWAMINATHAN, M. (with SUSHEELA, N.)

Supplementary value of edible coconut meal to poor Indian
diets based on rice, ragi, wheat and surghum. J. Nutr.
Diet. 1968   5: 104-109.

289 DANIEL, V.A., URS, T.S.R., DESAI, B.L.M., RAO, S.V.,
RAJALAKSHMI, D., SWAMINATHAN, M. and PARPIA, H.A.B.
Studies of low cost balanced foods suitable for feeding
weaned infants in developing countries. I. The protein ef-
ficiency ratio of low cost balanced foods based on ragi, or
maize, ground nut, Bengal gram, soya and sesame flours
and fortified with limited amino acid. J. Nutr. Diet.
1967   4:183-188.

290 DARLING, H.S.
Annual Report of the Agricultural Entomologist (1944-45).
Rep. Dep. Agric. Uganda. 1944-45, 2:25-30.   Entebbe,
1946.

291 DASTANE, N.G., JOSHI, M.S. and SINGLACHAR, M.A.
How to prevent waterlogging.   Indian Fmg. 1963.   13:
No. 3:7-8.

292 DAWSON, J.H. and BRUNS, V.F.
Emergency of barnyard grass, green foxtail and yellow
foxtail seedlings from various soil depths.   Weeds 10(2):
136-139.   1962.

293 DAWSON, M.J.
Effect of seed-soaking on the growth and development of
crop plants. I. Finger millet (Eleusine coracana Gaertn.)
Indian J. Plant Physiol. 8(1): 52-60.   1965.

294 DEBRE ZEIT BRANCH EXPERIMENT STATION, ETHIOPIA
A progress report on cereal and oil seed research 1955-
1963.   Experiment Station Bulletin 39.   Imperial Ethi-
opian College of Agricultural and Mechanical Arts 1965
pp. 60.   (Bibl. 7) received Nov. 1967).

295 De CANDOLLE, A.
Origin of the cultivated plants.   Kegan Paul, Trench and
Co.   London 1886.

296 de CUNTO, M.
Estudo sobre o adlay (Coix lachryma-jobi, de Linen).   Rev.
Nutricao 1950, 1:47-65.

297 DEIGHTON, F.C.
Mycological Section - 1927.   Annual Report of the Lands
and Forest Department.   Sierra Leone for the year, 1927,
pp. 13-17.   1928.

298 DEIMAN, A.I.
Scientific report of the Alma-ata State Breeding Station

(millet breeding). Ogiz, Selkhozguz, Moscow 1945; pp. 69-73.

299 DELASSUS, M.
The main diseases of millet and sorghum observed in Upper Volta in 1963. Agron. Trop., Paris 1964, 19:489-498.

300 DELBOSC, G.
A study on the regeneration of soil fertility in the ground nut zone of Senegal. Oleagineux 1968, 23, No. 1, 27-33 (Bibl. I. F. Eng. Span. Stn IRHO Darou Senegal).

301 DENISENKO, Ya I. and BELOVA, S. M.
Hydrogenation of millet oil. Prkl. Biokhim. Mikrobiol. 2(2): 218-222. 1966.

302 DEO, R., BASER, B. L., RUHAL, D. V. S.
Effect of sodium salts on the growth and mineral composition of bajra (Pennisetum typhoides). Ann. Arid Zone, 1968, 7, No. 2, 100-104. (Bibl. 10, Agr. Exp. Stn., Univ. Udaipur, Rajadthan, India).

303 DEO, V. R.
Study of Paspalum scrobiculatum, extract in forty psychotic patients. Psychopharmacologia 5(3):228-233. 1964.

304 DEPARTMENT OF AGRICULTURE, NYASALAND PROTECTORATE
Finger millet. Method of planting trial at Baka, Karonga. Rep. Dep. Agric. Nyasald 1959-60. Pt. 2, 1961, 29-30.

305 De PAULA SANTOS, C.
Contribuicao para o estudo do valor nutrivo do adlay (Coix lachryma-jobi, L.). An. Fac. Med. Univ. Sao Paulo, 1950-1951: 25: 323-342.

306 DERJABINA, A. P.
Promising millet varieties. Trans. Orenburg Agric. Inst. 1962, 11:120-26: from Ref. Z. (Ref. J.) 1963, Abst. 9. 55.135.

307 DERJABINA, A. P. and JAGER, F. H.
The Orenburg 42 millet. Selek. Semenovodstvo 1965, No. 4:66-67.

308 DESAI, M. T. and PATEL, R. M.
Some observations on the biology and control of white grubs in soil (Holotrichia near Consanguinea Blanch,) affecting groundnut and cereals in Gujarat. Indian J. Ent. 27(1):89-94. N. Delhi, 1964.

309 DESAI, S. G., DESAI, M. V., and PATEL, M. K.
     Control of bacterial blight of ragi (Eleusine coracana) by
     Streptocyclic. Indian Phytopathyol. 20(4): 294-295. 1967
     (Rec'd 1968).

310 DESAI, S. G., THIRUMALACHAP, M. J. and PATEL, M. K.
     Bacterial blight disease of Eleusine coracana Gaertn.
     Indian Phytopath. 18(4): 384-386. 1965.

311 DESCAMPS, M.
     Comportement du criquet migrateur africain (Locusta
     migratoria migratorioides Rch. and Frm.) en 1957 dans
     la partie septentrionale de sonaire de gregarisation sur
     le Niger. Region de Niafunre (Behaviour of L. M. mi-
     gratorioides in 1957 in the northern part of its outbreak
     area on the Niger. Report of its outbreak area on the
     Niger. Region of Niafunke). Locusta No. 8, 280 pp.
     Locusta No. 8 280 pp Nogent-sur-Marne, 1961.

312 DESIKACHAR, H. S. R., and DE, S. S.
     The cystine and methionine contents of common Indian
     foodstuffs. Curr. Sci. 16:284. 1947.

313 DEY, P. K.
     Administrative Report on the Department of Agriculture,
     Uttar Pradesh, 1947-48. 1949.

314 DICKSON, J. G.
     Diseases of field crops. 2nd Ed. New York: McGraw-
     Hill. pp. 1-517. 1956.

315 DILLMAN, A. C.
     Breeding millet and sorgho for drought adaptation. USDA
     Agr. Bull. 291. 1961.

316 DILLMAN, A. C.
     The water requirements of certain crop plants and weeds
     in the northern Great Plains. J. Agr. Res. 42(4): 187-
     238. 1931.

317 DIVAKARAN, K.
     A new species of Eleusine (Gaertn). Madras Agr. J.
     1959, 46:485-86.

318 DIVAKARAN, K.
     Studies on age of seedlings in ragi - Eleusine coracana
     Gaertn. Madras Agr. J. 1967 54. No. 9. 658-664.
     (Bibl. 7; Agric. Coll. and Res. Inst. Coimbatore, India.

319 DIVAKARAN, K., SURENDRAN, C. and GOVINDASAMY, K. N.
     A note on ratoon cropping of 'kodra" (Varagu), Paspalum
     scrobiculatum Linn. Madras Agr. J. 1966, 53(10): 425-
     426. 1966.

320 DJADJUN, P. M.
The Novo-Urensk State Breeding Station. Selek. Semenovodstvo 1950, No. 25-29.

321. DMITRENKO, P. A. and SHILURMOVA, V. S.
Dependence of availability of phosphate on the time of
manuring and the conditions of reaction of fertilizers with
the soil. Dokl. Akad. Nauk. SSSR 76, 447-450. 1951.

322 DOI, Y. and KIMURA, M.
Observations on protoplasmic streaming in the cells of
crop plants. 1. Protoplasmic streaming in various hair
cells. Bull. Fac. Agric. Yamaguchi 1959, No. 10: 1225-
1234, Bibl. 14.

323 DOIDGE, E. M.
The South African fungi and lichens to the end of 1945.
Bothalia 5:105. 1950.

324 DONEGHUE, R. C.
Crop rotation and soil fertility. N. Dak. Sta. Bull. 126
(1918), pp. 197-251. 1918.

325 DORNBURG (GERMANY), ABSCHLUSSBER. INST. PFLAN-
ZEN
Mutations versuch zue Gewinnung einer fruhsaatvertrag-
lichen hirse mit rascher jugendentwicklung (Mutation experiment to produce a millet tolerant of early sowing and
having rapid juvenile development). Abschlussber. Inst.
Pflanzens. Dornburg/Saale Friedrich-Schiller Univ. Plan-
Nr. 2156251-12/7:pp. 7; from Landw. Zbl.: Abt. II(8):
2482. 1963.

326 DOTZENKO, A. D., COOPER, C. S., DOBRENZ, A. K., LAV-
DE, H. M., MASSENGALE, M. A. and FELTNER, K. C.
Temperature stress on growth and seed characteristics of
grasses and legumes. Tech. Bull. 97. Colo. Agric. Exp.
Sta. 1967 pp. 27. (Bibl. 13).

327 DOUGLAS, N. J.
Early summer feed for dairy cows in South Coastal
Queensland. Qd. Agric. J. 85(10):632-635. 1959.

328 DOVRAT, A. and OPHIR, N.
The effect of number of cuttings, seeding rate and row
spacing on yield and leaf area index of pearl millet (Pennisetum glaucum L. R. Br.) Israel J. Agric. Res. 15:
179-186. 1965.

329 DOWNS, R. J., BORTHWICK, H. A. and PIRINGER, A. A.
Comparison of incandescent and flourescent lamps for
lengthening photoperiods. Proc. Amer. Soc. Hort. Sci.
71: 568-578. 1958.

330 DRECHSLER, C.
Some graminicolous species of Helminthosporium. J. Agr.
Res. 24:641-739. 1923.

331 DUBYANSKY, V.A.
Utilization of virgin and waste sandy soils of the Don area
sands. Zemledelie 1954 (10):12-20. 1954. Referat. Zh.
Biol., 1956, No. 13802.

332 DUDKIN, M.S. and MEDVED'EVA, E.I.
A study of the xylans of oat, barley and millet hulls. In:
Uglevody i uglevodnyi Obmen. (Carbohydrates and Carbo-
hydrate Metabolism). Izo. Akad. Nauk. S.S.S.R.: Moscow. 49-55. 1962.

333 DUMONT, S.
Millets and sorghums grown in the East of the Republic
of Niger. Agron. Trop. 21(8):883-917. 1966.

334 DUNBAR, A.R.
The annual crops of Uganda. East African Literature
Bureau, Dar es Salaam, Nairobi, Kampala: 1969.

335 DUPRIEZ, G.I.
Evaporation and water requirements of different crops in
the Muuazi Region (Lower Congo). Publ. Inst. Nat. Agron.
Congo. Ser. Sci. 1964, 106: pp. 106.

336 DUTHIE, J.F. and FULLER, J.B.
The fodder grasses of Northern India. Rourkee. 1888.

337 DZHUMAGULOVA, L.I. and RAKHIMBAEV, I.
Alteration of some biochemical properties of millet grains
on drying and the influence of temperature and moisture
content. Izv. Akad. Nauk. Kaz. SSR. Ser. Biol. 4(3):
57-61. 1961.

338 EAST AFRICAN..........E.A.A.F.R.O. Annual Report 1955.

339 EHARA, K. and ABE, S.
Studies on Echinochloa crusgalli as a weed in rice fields.
VI. Classification of the types of E. crusgalli occurring
in rice fields. (Proc. Crop Sci. Soc. Japan, 1952: 20;
245-246.

340 EL-KHALAFY, H.M., SHOEB, Z.E. and GAD, A.M.
Chemical investigations on Egyptian vegetable fats and oils.
The chemical constitution of some Gramineae seed oils
Grasas Aceites 18/67 291-295. Ill. 1967 (Sp., Eng. &
Fr. sum.).

341 ELLIOTT, C.
     A bacterial stripe disease of proso millet. J. Agr. Res.
     26:151-160. 1923.

342 ELLIOTT, C.
     Manual of bacterial plant pathogens. Chronica Botanica
     Co., Waltham. 1951.

343 ELLIOTT, C. and POOS, F.W.
     Seasonal development, insect vectors, and host range of
     bacterial wilt of sweet corn. J. Agr. Res. 60(10): 645-
     686. Wash., D.C. 1940.

344 ELSUKOV, M. P.
     An experiment on changing the heritable basis of some
     annual forage plants. Selek. Semenovodstvo 1956: No 4:
     36-44.

345 EMERSON, R. A.
     Cover crops for young orchards. Nebr. Sta. Bull. 92.
     pp. 1-23 pls 2. Exp. Sta. Rec. 1906-07.

346 EMBERETSON, A. B.
     Japanese barnyard millet: A new forage for the coast
     section. Oregon Sta. Circ. 80, pp. 4. 1927.

347 EMIKH, T. A.
     Controlling the development of plants. In: Fiziologiches-
     kaya Pitaniya, Rosta i ustoichivosti Rastenii v Sibiri i na
     Dal'nem Vostoke (Physiological Nutrition, Growth and Re-
     sistance of Plants). Akad. Nauk SSSR: Moscow 19-23.
     1963.

348 EMIKH, T. A.
     Effect of a short day on the nucleic acids in the growing
     tips of millet stalks. In: Biologiya Nukleinogo Obmena u
     Rasteniyakh. Nauka: Moscow 152-157. 1964 (Biology of
     Nuclear Metabolism in Plants).

349 ERICKSON, D.O., HAUGSE, C.N., DINUSSON, W.E., BOLIN,
     D.W. and BUCHANAN, M. L.
     Prose: lysine deficient for pigs. N. Dak. Farm Res.
     1963: 22: No. 9:7-8.

350 ERMAKOV, A.I., YAROSH, N.P. and MIKHAILOV, A.A.
     O Kolichestvennom Opredelenii Triptofana v Semenakh.
     (Quantitative determination of tryptophan in seeds (barley,
     rice, millet, soft wheat, hard wheat, wild wheat species,
     rye). Prikl. Biokhim. Mikrobiol. 3(1): 107-112. 1967.

351 ESTIFEYEFF, P.G.
     Disease of cultivated and wild plants in the Djetyssouy
     Region in the period 1923-24. Pamph. Djetyssouy Pl.

Prot. Sta. Alma-Ata. pp. 14. (Rev. Appl. Mycol. 4:445).
1925.

352 ETASSE, C.
The improvement of Pennisetum millet in Senegal. Agron.
Trop., Paris 1965, No. 10:976-80.

353 ETASSE, C.
Improvement of Pennisetum millet at the Bambey C. R. A.
(Senegal). Sols Afr. 11(1 and 2):269-275. 1966.

354 EVANS, M.W. and GROVER, F.O.
Developmental morphology of the growing point of the shoot
and inflorescences in grasses. J. Agr. Res. 61:481-521.
1940.

355 FAIRBROTHERS, D.E.
Relationships in the Capillaria group of Panicum. Diss.
Abstr. 1954. 14: Publ. No. 9913:1898-1899.

356 FANOUS, Mamdouth, A.
Test for drought resistance in pearl millet (Pennisetum
typhoideum). Agron. J. 59(4):337-340. 1967.

357 FEDOROV, A.K.
The effect of different day lengths on the development of
millet. Tr. Inst. Genet. Akad. Nauk SSSR 30:129-135.
1963.

358 FEDOSEEVA, Z.M. and ZUBKO, I. Ya.
Rol' orosheniya v. uskorenii osvobozhdeniya pochvy ot in-
fektsii golovni prosa i yachmenya. (The role of irrigation in accelerating the removal of millet and barley smuts
from the soil.) Uchenye Zap. Kharkov. Univ., 141:109-
114. 1963. Referat. Zh. Rasten., 1964(17). 492. 1964.

359 FILLIPEV, I.D. and USTENKO, Z.
The productivity of sorghum and millet and the composition of locally applied fertilizers. Vestnik Sel-kh. Nauk.
1965. No. 6:42-44.

360 FISCHER, G.W.
Manual of the North American smut fungi. Ronald Press
Co., New York. 1953.

361 FLETCHER, T.B.
Annotated list of Indian crop pests. Bull. Agric. Res.
Inst. Pusa. 100: 1-246. 1921.

362 FLORIDA EXPERIMENT STATIONS
Annual Report for the Fiscal Year ending 30 June, 1962.

Univ. of Fla. Agr. Exp. Sta. Ann. Report 1962: pp. 370.

363 FOKEEV, P.M.
Cereal plants for cultivation under irrigation. Selek.
Semenovodstvo 1953: No. 1: 56-59.

364 FOOD and AGRICULTURAL ORGANIZATION (UN)
Report of the CCTA/FAO Symposium on storage of food
crops in Africa. Held in Freetown, Sierra Leone, 20-24
February, 1962.

365 FORD, D. and SCOTT, R.
The native economies of Nigeria. Faber and Faber, Ltd.,
24, Russell Square, London, pp. 333. 1946.

366 FORTMANN, H.R., CARNAHAN, H.L., PENNINGTON, R.P.
and WASHKO, J.B.
Performance of sudangrass varieties and millets at four
locations in Pennsylvania in 1949. Prog. Rep. Pa. Agric.
Exp. Sta. 1950, pp. 7.

367 FROMANTIN, Jane
Growth inhibiting action of extract from the integuments
of fruits of Lophira alata (Ochnaceae). Bull. Soc. His.
Natur. Toulouse 103/1/2 99-104. Illus. 1967.

368 FUKUI, H., MOTOYAMA, E. and KUBOTA, S.
An experiment on the basic exchange capacity and selec-
tive absorption of base, of forage crops. Bull. Shikoku
Agric. Exp. Stn. 1964. 10. 123-128.

369 FURLONG, J.R.
The estimation of hydrocyanic acid in feeding stuffs and
its occurrence in millet and Guinea corn. Analyst 39
(463):430-432. 1914.

370 GALVEZ, Guillermo E.
The incidence of Sogatei orizicola Muri and Sogatei urbana
Crauf. (Homopteia, Araepodidae) in plantations of rice and
in the barnyard grass (Echinochloa colonum) in Colombia.
Cent. Nac. Invest. Agropecuaria Tibaitata, Colombia.
Agric. Trop. 23(6): 384-398. Illus. 1967 (Eng. sum.).

371 GALVEZ, Guillermo E.
Transmission studies of the Hoja blanca virus, with highly
active, virus free colonies of Sogatodes orizicola. Phytopathology 58(6): 818-821. 1968 (Span. sum.). Inst. Col-
umbiano Agropecuaria, Bogota, Colombia.

372 GAMBIA, Department of Agriculture
Annual Report of the Department of Agriculture, Colony

of Gambia, for the year ending 31 May, 1942.

373 GAMBLE, J. S.
    Flora of the presidency of Madras, Pt. 10-Gramineae,
    London, 1934.

374 GAMMIE, G. A.
    Field and garden crops of the Bombay presidency.   Bull.
    Dep. Agric. Bombay, 30.   1908.

375 GANAPATHY, N. S., CHITRE, R. G. and GOKHALE, S. K.
    The effect of the protein of Italian millet (Setaria italica)
    on nitrogen retention in albino rats.   Indian J. Med. Res.
    1957, 45: 395-399.

376 GANAPATI, S. and CHITRE, R. G.
    A study on enzymatic liberation (in vitro) of essential
    amino acids and proteins of Italian millet (Setaria italica)
    and field bean Dolichos lablab).   J. Postgrad. Med. 1961.
    7: 158-163.

377 GANAPATI, S. and CHITRE, R. G.
    The effect of Italian millet (Setaria italica) and field bean
    (Dolichos lablab) proteins on the formation of hemoglobin
    and plasma proteins in albino rats.   J. Postgrad. Med.
    1961, 7: 164-166.

378 GANCHEFF, J.
    Composition and feeding value of millet straw and hay of
    vetch grown with oats.   Ann. Univ. Sofia. Fac. Agric.
    6:1-12.   1928.

379 GANGSTAD, E. O.
    Pearl millet and grass sorghum.   Grass Mimeograph No.
    19.   Renner, Texas, 1959.

380 GARKAVYJ, P. F. and DANIL'CUK, P. V.
    Agrobiological evaluation of some Setaria varieties and
    forms under the conditions of the southern Ukraine.   Proc.
    Lenin Acad. Agric. Sci. 1958. No. 2:-38.

381 GARRISON, W. D.
    Forage crops grown at coast land experiment station.
    South Car. Sta. Bull. 123, pp. 15.   1907.

382 GARWOOD, E. A.
    Studies on the root development of grasses.   Exp. Prog.
    Grassl. Res. Inst. Hurley 1960-61, 15th report. 1962,
    44-5.

383 GAST, M. and ADRIAN, J.
    Mils et sorgho en ahaggar.   Etude ethnologique et nutri-
    tionelle.   Mem.   Centre Recherches Anthropol.   Prehis-

toriques Ethnograph., Algiers, 1965, No. 4, pp. 77.

384 GEERING, Q. A.
A cotton stainer (Dysdercus superstituosus Fabr.) as a potential pest of sorghum. Empire J. Exp. Agric. 20 (79): 234-239. Oxford, 1952: also in Res. Mem. Emp. Cott. Gr. Corp. No. 17. London, 1953.

385 GEERING, Q. A.
The sorghum midge, Contarinia sorghicola (Coq.) in East Africa. Bull. Ent. Res. 44(2): 363-366. London, 1953.

386 GEIGER, M. and BARRENTINE, B. F.
Isolation of the active principles in Claviceps paspali - A progress report. J. Amer. Chem. Soc. 16:966-967. 1939.

387 GEISSEN PLANT BREEDING INSTITUTE
Novelties for the trade. Saatgut-Wirt. 1962. 14:126-127.

388 GENERALOV, G. and PETROVA, Z.
New varieties of gruel crops. Kolkhoz. Proizv. 1960. No. 9:39-40.

389 GEORGE, L.
The andat bug or gutran el andat. Sudan Notes and Records 37:96-98. 1956 (1958).

390 GEORGIA (UNIV) COLLEGE OF AGRICULTURE, AGRICULTURE EXPERIMENT STATIONS
Serving Georgia through Research, 1964 Annual Report. Ga. Agric. Exp. Sta. Ann. Rpt. 1964. pp. 90.

391 GHOSH, A. K. and others
Report of the Agronomy Department of the Agricultural Institute, Allahabad. Allahabad Fmr. 29(5):115-141. 1955.

392 GILDENHUYS, P. J.
Fertility studies in Setaria sphacelata (Schum.) Stapf and Hubb. Sci. Bull. Dep. Agric. S. Afr. 1950, No. 314: pp. 45.

393 GILDENHUYS, P. J. and BRIX, K.
Apomixis in Pennisetum dubium. S. Afr. J. Agric. Sci. 1959, 2:231-245.

394 GILDENHUYS, P. and BRIX, K.
Genic control of aneuploidy in Pennisetum. Heredity 1961, 16:358-363.

395 GILDENHUYS, P. and BRIX, K.
Genically controlled variability of chromosome number in Pennisetum hybrids. Heredity 1964, 19:533-42.

396 GILDENHUYS, P. and BRIX, K.
     Relationship between embryo and endosperm in inter-spe-
     cific hybrids in Pennisetum (L.) Rich. Ann. Bot. 1965,
     29:709-715.

397 GILE, P. L.
     Effect of different soil colloids on the toxicity of boric
     acid to foxtail millet and wheat. J. Agr. Res. 70(10):
     339-346. 1945.

398 GILES, P. H.
     The storage of cereals by farmers in northern Nigeria.
     Trop. Agric. Trin. 41(3): 197-212. London, 1964.

399 GILES, P. H.
     The storage of cereals by farmers in northern Nigeria.
     Samaru Res. Bull. 62:197-212. 1965.

400 GILL, B. S. and GUPTA, A. K.
     Study of karyomorphology in Pennisetum typhoides Stapf
     and Hubb. J. of Res. (India) Vol. III(2):118-121. June,
     1966.

401 GILL, B. S., SRAON, H. S. and MINOCHA, J. L.
     Colchicine induced tetrapolidy in Pennisetum typhoides
     Stapf and Hubb. J. of Res. (India) Vol. III(3):260-263.
     Sept. 1966.

402 GILL, B. S., GUPTA, V. P. and NAGI, K. S.
     Inheritance of some quantitative characters in pearl mil-
     let. J. Res. (India) 1968, 5: No. 2:37-40.

403 GIRI, K. V.
     The availability of calcium and phosphorus in cereals.
     Indian J. Med. Res. 28(1): 101-112. 1940.

404 GODBOLE, S. V.
     Seasonal influence on the water requirement and growth of
     bajri (P. typhoideum). Proc. 13th Indian Sci. Congr. 211.
     1926.

405 GODBOLE, S. V.
     Pennisetum typhoideum studies on the bajri crop. I. The
     morphology of Pennisetum typhoideum. Mem. Dep. Agric.
     India Bot. Ser. 14:247-268. 1927.

406 GOEL, L. B., MATHUR, S. B., JOSHI, L. M.
     Seedborne infection of Piricularia setariae in Setaria
     italica. Pl. Dis. Reptr. 51(2) 1938. 1967.

407 GOKHALE, V. C.
     Rep. Dep. Agric. Bombay, 1936-37: 188. 1937.

408 GOLOVTSEV, L. A.
Vegetative hybridization of cereals. Agrobiologiya no. 1,
p. 153-157. Jan./Feb. 1948.

409 GONCAROV, P. L.
Growing Setaria in short days as a means of obtaining
early varieties. Vestnik Sel-kh. Nauk. 1963, No. 8(3).
113-115.

410 GONCAROV, P. L.
On the methods of breeding agricultural crops. Selek.
Semenovodstvo 31(2):48-53. 1966.

411 GOODEARL, G. P.
Millet feeding for turkeys. N. Dak. Sta. Bimo. Bul.
1(5):3-6. 1939.

412 GOUWS, L. and KISTNER, A.
Bacteria of the ovine rumen. 4. Effect of change of diet
on the predominant type of cellulose-digesting bacteria.
J. Agric. Sci. 1965. 64:51-57.

413 GOVANDE, G. K.
New strains of pulses and millets in Baroda State. Indian
Fmg. 1950. 11: 153-154.

414 GOVINDA-RAJAN, S. V. and GOPALA-RAO, H. G.
Effect of micronutrients on crop response and quality in
Mysore State. J. Indian Soc. Soil Sci. 1964, 12: 355-
361.

415 GOVINDU, H. C., RENFRO, B. L., DIXIT, L. A. and SKOLD,
L. N.
The reaction of the Indian collection of Eleusine coracana
to Pellicularia rolfsii/nematode disease complex. Indian
Phytopathol. 19:126-127. 1966.

416 GOVINDU, H. C. and SHIVAIVANDAPPA, N.
Studies on an epiphytotic ragi (Eleusine coracana) disease
in Mysore State: (Agric. Coll. Univ. Agr. Sci. Bangalore, Mysore, India. Mysore J. Agric. Sci. 1(2): 142-
148. Illus. 1967.

417 GOYAL, R. D.
A 'spreading' mutant in bajra. (Pennisetum typhoides, S.
& H.) Sci. Cult. 1962, 28:437-438.

418 GRABOUSKI, P.
New proso millet developed for short-season replacement
crop. Univ. of Nebr. Quarterly XI (3):9. Fall, 1967.

419 GRANITI, A.
Su alcuni fungilli graminicoli dell'Africa orientale (some

graminicolous fungi of East Africa). Nuovo Gior. Bot.
Ital., N.S. 57(1-2): 247-56. 1950.

420 GRASSLANDS RESEARCH STATION, MARANDELLAS
A study of effects of fertilizers and planting-density on
the yield of rapoke (Eleusine corcana). Rep. Grassl. Res.
Sta. Marandellas 1956-7, 21-22.

421 GRASSO, V.
Le Claviceps delle graminacee Italiane. Ann. Sper. Agr.
(n. s.) pp. 143. 1952.

422 GREB, B.W.
Per cent soil cover by six vegetative mulches. Agron J.
1967:59: No. 6: 610-611 (Bibl. 2. USDA. Akvon. Colorado).

423 GREEN, V.E. and ORSENIGO, J.R.
Wild grasses as possible alternate hosts of hoja blanca
(white leaf) disease of rice. Pl. Dis. Reptr. 42(3):342-
345. 1958.

424 GREWAL, J.S. and PAL, M.
Seed microflora I. Seedborne fungi of ragi (Eleusine
coracana Gaertn.) their distribution and control. Indian
Phytopathol. 1965. 18:33-37.

425 GREWAL, J.S. and PAL, Mahendra
Seed microflora II. Seedborne fungi of Setaria italica,
their distribution and control. Indian Phytopathol. 1965.
18:123-27.

426 GRICENKO, G.D. (GRITSENKO, G.D.)
The powerful force of the cultivar. Selek. Semenovodstvo
No. 31:9-14. 1966.

427 GRIFFING, B.
A generalized treatment of the use of diallel cross in
quantitative inheritance. Heredity 10:31-50. 1956(a).

428 GRIFFING, B.
Concept of general and specific combining ability in relation to diallel crossing systems. Aust. J. Biol. Sci. 9:
463-493. 1956(b).

429 GRISHKEVINCH, V.A.
Vliyanie standiingo sostoyaniya prosa na kolichestvennyi
sostav rizosfernykh bakteaii (The effect of the development
stage of proso millet on the quantitative composition of
rhizosphere bacteria). Zap. Leningradsk. Sel'skokhoz.
Inst. 13.103-107. 1958: Referat. Zh. Biol., 1960, No.
5861.

430 GRODZINSKII, A. M.
Competitive inter-relations between millet (Setaria glauca)
and weeds. Ukr. Bot. Zh. 1960, 17: No. 6:54-57.

431 GROZIER, A. A.
Millet. Mich. Sta. Bull. 117, p. 64. Exp. Sta. Rec. 6:
713. 1894-1895.

432 GUERNIER, M.
Rice cultivation without irrigation in casamance (South
Senegal). Riz et Rizicult. 1 (4):131-133. 1955.

433 GUJARAT AGRIC. DEPT.
Hybrid Bajri in Gujarat. Aryaswapatra 3(11):3. 1967.

434 GUNTHER, E. and MESCHKAT, Hilde
The importance of trace elements and their influence on
the microbial activity of the soil and the growth of early
stages of higher plants. Dtsch. Landw. 1955-6. No. 10:
498-501.

435 GUPTA, V. P. and ATHWAL, D. S.
Genetic variability, correlation and selection indices of
grain characters in pearl millet. J. Res. (India) Vol. III
(2): 111-117. 1966.

436 GUPTA, V. P. and ATHWAL, D. S.
Genetic variability, correlation and selection indices for
green fodder characters in pearl millet. J. of Res.
(India) Vol. III (4): 379-384. Dec. 1966.

437 GUPTA, V. P. and RACHIE, K. O.
Importance and utilization of millets in India's agriculture.
J. Postgrad. Sch. IARI 2:1. 1964.

438 GUPTA, V. P. and SINGH, H.
Role of genetic diversity in combining ability of Pennisetum inbreds. J. Res. (India) Ludhiana 4:41-46. 1967.

439 GUPTA, V. P. and NANDA, G. S.
Inheritance of grain characters in pearl millet. J. Res.
(India) 1967. 4:488-491.

440 GUPTA, V. P. and NANDA, G. S.
Inheritance of some plant and head characters in pearl
millet. J. Res. (India) 1968. 5(1): 1-5.

441 GURCHARAN-SINGH, S. and SINGH, S. Ranjit
Bajra hybrid I effects a revolution. Indian Fmg. 16(2):
12-13. 1966.

442 GUSEIN, I. V.
Varieties of field crops of the Krasno-Kut State Selection

Station. (35 years' work of the station.) Soc. Zern. Hoz.
1946, No. 4:35-40.

443 GUTTIKAR, Mankernika N., MANGALORE, Myna Pane, JAYA-
RAJ, A. Paul, RAO, M. Narayana, and SWAMINATHAN, M.
Studies on processed protein foods based on blends of
groundnut, Bengal gram, soya bean and sesame flours and
fortified with minerals and vitamins.
IV. Supplementary value to diets based on ragi (Eleusine
coracana) on maize and tapioca.
Cent. Food Technol. Res. Inst., Mysore, India. J. Nutr.
Diet. 5(2):110-120, 1968.

444 GYSIN, H.
Chemical constitution and selective effect of triazine herbicides.   Pflanzenschutzberichte 38(12):211-223.   Illus.
1968.

445 HACKEL, E.
In Engler and Prantl, "Die Naturlichen Pflanzen Familien,
II Graminere." Leipzig. W. Engelman. 1887.

446 HACKER, J. B.
The maintenance of strain purity in the Setaria sphacelata
complex. J. Aust. Inst. Agric. Sci. 1967. 33:265-67.

447 HADJIMARKOS, D. N.
Selenium content of millet and dental caries in the rat.
Nature 1962, 193:178.

448 HAENSELER, C. H.
Helminthosporium leaf spot on millet in New Jersey. Pl.
Dis. Reptr. 25:486. 1941.

449 HANKS, R. J., GARDNER, H. R. and FLORIAN, R. L.
Evapotranspiration - climate relations for several crops
in the Central Great Plains. Agron. J. 60(5): 538-542.
Illus. 1968.

450 HANSEN, N. E.
Proso and Kaoliang as table foods. S. Dak. Agric. Exp.
Sta. Bull. 158. 1915.

451 HANSFORD, C. G.
Contribution towards the fungus flora of Uganda. V.
Fungi Imperfecti. Proc. Linn. Soc. London. 1942-
1943. pp. 34-67. 1943.

452 HANSFORD, C. G.
Report of the Department of Agriculture of Uganda. Rep.
Dep. Agric. Uganda p. 48-57. 1933.

453 HARADA, K., MURAKAMI, M., FUKUSHIMA, A. and NAKA-
ZIMA, M.
Studies on the breeding of forage crops. I. Studies on in-
tergeneric hybridization between Zea and Coix. Sci. Rep.
Saikyo Univ., Agric. 1954, No. 6: 134-145.

454 HARGREAVES, H.
Report of the Senior Entomologist (1938-39). Rep. Dep.
Agric. Uganda 1938-1939, 2:5-9. Entebbe, 1940.

455 HARIHARAN, K., RAJURS, T.S.S., DESAI, B.L.M., RAO,
S.V., RAJALAKSHMI, D., SWAMIMATHAN, M., and PARPIA,
H.A.B.
Effect of supplementary poor Indian diets based on Kaffir
corn, pearl millet, and maize with vitamins and minerals
and fortified/unfortified groundnut flour on the nutritive
value of the diets as judged by the growth of rats and on
the protein ratio. J. Nutr. Diet. 1965. 2: 196-201.

456 HARRIS, E.
Distortion of Guinea corn (Sorghum vulgare) caused by a
mealybug, Heterococcus nigeriensis. Williams, in Northern Nigeria. Bull. Ent. Res. 51(4): 477-684. London,
1061.

457 HARRIS, W.V.
Annual Report of the Entomologist for the year 1942.
Department of Agriculture, Tanganyika (Morogoro), 1943.

458 HART, R.H.
Digestibility, morphology and chemical composition of
pearl millet. Crop. Sci. 7(6): 581-584. 1967.

459 HART, R.H. and BURTON, G.W.
Effect of row spacing, seeding rate, and nitrogen fertiliza-
tion on forage yield and quality of Gahi-1 pearl millet.
Agron. J. 57:376-378. 1965.

460 HARTLEY, W.
Studies on the origin, evolution and distribution of the
gramineae, II. The tribe Paniceae. Aust. J. Bot. 6:
343-357.

461 HARTLEY, W. and SLATER, C.
Studies on the origin, evolution and distribution of the
Gramineae. III. The tribes of the subfamily Eragrostidae.
Aust. J. Bot. 8(3):256-276. 1960.

462 HARTWELL, B.L. and PEMBER, F.R.
The feeding power of certain cereals and their response to
fertilizer ingredients. Rhode Island Sta. Bul. 190. pp.
4-27. 1922.

463 HASELER, E.R.
Broom millet. Qd. Agric. J. 59:197-201. 1944.

464 HASHIOKA, Y. and ANDO
Phytopharmocology of rice diseases. Höfchenbr. Bayer
PflSchutz-Nachr. (Eng. Ed.) 4:177-190. 1955.

465 HAUGSE, C.N., DINUSSON, W.E., ERICKSON, D.O., BOLIN,
D.W. and BUCHANAN, M.L.
Proso for swine: levels of protein and lysine supplemen-
tation. N. Dak. Farm Res. 1962, 22:No. 8:15-18.

466 HAYES, J.F.
Experimental milling tests of Adlay. Philipp. Agric. Rev.
16(3):197-200. 1923.

467 HAYLETT, D.G.
Green manuring and soil fertility. S. Afr. J. Agric. Sci.
1961. 4: No. 3: 363-378.

468 HAYMAN, B.I.
The theory and analysis of diallel crosses. III. Genetics
45:155-172. 1960.

469 HAYWOOD, A.H.
Manchurian millets. Agric. Gaz. N.S. W. 21(9): 735-745.
1910.

470 HECKE, L.
Experiments in combating millet smut. Ztschr. Landw.
Versuchsw. Osterr., S. (1902), No. 1, pp. 22-28. Exp.
Sta. Rec. 14:457, 978. 1902-03.

471 HENRARD, J. Th.
Monograph of the genus Digitaria. Universitaire Pers Leiden. Leiden. 1950. pp.237, 238, 239, 338, 339.

472 HILL, A.G.
The improvement of native food crops. A precis of the
more important work done in East Africa during 1944 (millets). E. Afr. Agric. Res. Inst. Amani 1945, No. DF/
5/2: pp. 15.

473 HINEBAUCH, T.D.
Feeding of millet to horses. N. Dak. Agr. Exp. Sta.
Bul. 26. 1896.

474 HIRAYAMA, S., YAMAMOTO, M., SAKABE, F. and KOTO, K.
Experiments in Coix smut control. Bull. Natl. Hyg. Lab.
(Tokyo) 72:219-224. 1954.

475 HIROE, I.
Brachysporiosis of Plants. IV. Ann. Phytopath. Soc.

5:121-144. 1935.

476 HIROE, I.
Three new blight diseases of certain plants of Gramineae
and Cyperaceae. Ann. Phytopath. Soc. Japan 5:318-335.
1936.

477 HIRSCHHORN, E.
Note on synonymy. Tolyposporium senegalense is a syn-
onym of T. bullatum. Rev. Argent. Agron. 8:384-386.
(cf. Rev. Appl. Mycol. 24:62.)

478 HITCHCOCK, A.S.
Manual of the grasses of the United States. USDA Misc.
Publ. No. 200. Wash. 1950.

479 HIURA, M.
A simple method for the germination of oospores of Sclerospora graminicola. Sci. N.S. 72(1896):95. 1930.

480 HOCKINGS, E.T. and VEITCH, R.
The Queensland Agricultural and Pastoral Handbook. Vol.
1. Farm crops and pastures. Queensland Dep. Agric.
Stock, Brisbane, 1902, 2nd Ed. pp. 330.

481 HOFFMAN, O.L. and others
Factors affecting the activity of 4-chloro-2-butynl N-(3-
chlorophenyl) carbamate as a selective herbicide. Weeds
8(2): 198-203. 1960.

482 HOOF, H.V. Van
The delphacid Sogata cubana, vector of a virus of Echinochloa colonum. Tijdschr. Planteziekten 65(5): 188-189.
1959.

483 HOOKER, J.D.
"Flora of British India." Reeve & Co., London, 1897.

484 HOOVER, A.A. and JAYASURIYA, G.C.N.
Microbiological assay of vitamins. 2. Riboflavin. Ceylon J. Med. Sci. 8:183-189. 1951-53.

485 HOSINO, Tetu and TUTUMI, Tatuzi
Time and order of flower opening in S. italica. Notes
from Agric. Exp. Sta. Korea 6:443-454. 1926.

486 HOVELAND, C.S.
Millet and sudan varieties for forage in Alabama. Var.
Rep. Alabama Agric. Exp. St. 1961 (April). pp. 4.

487 HOVELAND, C.S., ANTHONY, W.B. and SCARSBROOKE,
C.E.
Effect of management on yield and quality of Sudan Sor-

ghum-Sudan hybrid and Gahi-1 pearl millet.   Leaft. 76
Alabama Agric. Exp. Sta. 1967. pp. 7 (Auburn).

488 HRISTOFORON, I.
     Agrobotanical research on millet in Bulgaria.   Izv. Kamar.
     Narod. Kult. 1:225-238.   Sofia, 1946.

489 HSIN, Y. -F., SUN, C. -J., PENG, Y. -f, JIN, R. -H. and
     CHOU, E. Y.
     Chemical control of European corn borer in Liaoning.
     Acta Phytophylac. Sin. 3(3):299-306. 1964.

490 HSU, C. -C.
     The classification of Panicum (Gramineae) and its allies,
     with special reference to the characters of lodicule, style-
     base and lemma. J. Fac. Sci. Tokyo Univ.: Sect. III
     1965, 9:43-150.

491 HUME, A. N. and CHAMPLIN
     "Trials with millets and sorghums for grain and hay in
     South Dakota," S. Dak. Agric. Exp. Sta. Bull. 135. 1912.

492 HUNG-JU, Liang and KAI-SHENG, Liang
     The adaptability and the future of "Ta-Hung-P'ao" millet
     in Northern Shensi. Acta Agr. Sin. 10(4):318-326.   Aug-
     ust 1959.

493 HURSH, C. R.
     Social aspects of land use. J. Soil Wat. Conserv. Vol.
     21(6): 217-218. Nov. -Dec. 1966.

494 HUSAIN, A. and THAKUR, R. N.
     A new technique of inoculating pearl millet with Tolypos-
     porium penicillariae. Sci. Cult. 29(12): 607-608. 1963.

495 HUTSON, J. C.
     Report of the work of the Entomological Division, 1939.
     Ceylon Adm. Rep. Agric. 1939, D19-D20. Colombo, 1941.

496 IL'IN, V. I.
     Ways and methods of millet breeding in the Southeast.
     Selek. Semenovodstvo 1960. No. 2:54-57.

497 (INDIAN) ACAD. of AGRIC. SCI.
     Abstracts of papers presented for the Fifth Annual Session
     (1965), Madras Agr. J. 1965, 52:10. 445-459.

498 IARI (INDIAN AGRIC.....)
     Sci. Rep. Indian Agric. Res. Inst. 1953-54 pp. 87-95.
     1955.

499 IARI
Scientific report for the year ending 30 June, 1958. Sci.
Rep. Indian Agric. Res. Inst. 1958. pp. 212.

500 IARI
Agric. Res. (India) 1963, 3:1-14, 89-100, 163-176, 249-265.

501 IARI
The Open-pollinated cv. Pusa Moti. J. Postgrad. Sch.
IARI 1966. 4:No. 1-2. 168-176.

502 ICAR (Indian Council.....)
Annual Report of the Indian Council of Agricultural Research for 1950-51: pp. 170.

503 ICAR
Annual Report of the Indian Council of Agricultural Research for 1954-1955. pp. 175.

504 ICAR
"Special millets number." Poona Agric. Coll. Mag.
1955. 46:69-259. (Report of the All-India Millet Workers
Conference held for the first time at Kolhapur, Bombay
State, 16-18 May, 1955.)

505 ICAR
Agricultural Research Report. Rep. Indian Coun. Agric.
Res. 1958-1959, pp. 3-51. 1961.

506 ICAR
New Bajras. Indian Fmg. 1960, 10: No. 2: p. 14.

507 ICAR
"Handbook of Agriculture" Technical Editors: Kalidas
Sawhney and J.A. Da'ji; Chief Editor for Production: D.
Raghavan, Job Press Pvt. Ltd., Kanpur (1961).

508 ICAR
Progress Report of the Accelerated hybrid sorghum and
millet improvement programme 1963-1964. Indian Council of Agricultural Research and Co-operating Agencies,
New Delhi. 1964.

509 ICAR
Progress Report of the Co-ordinated Millets Improvement
Programme 1964-1965. Indian Council of Agricultural Research and Co-operating Agencies, New Delhi. 1965.

510 ICAR
Progress Report of the Co-ordinated Millets Improvement
Programme 1965-1966. Indian Council of Agricultural Research and Co-operating Agencies, New Delhi. 1966.

511 ICAR
Progress Report of the Co-ordinated Millets Improvement
Programme 1966-1967. Indian Council of Agricultural Re-
search and Co-operating Agencies, New Delhi. 1967.

512 ICAR
Progress Report of the All-India Co-ordinated Millet Improvement Programme 1967-1968. Indian Council of Agricultural Research and Co-operating Agencies, New Delhi,
1968.

513 INDIAN FARMING
Cereals: Bajra in arid areas.   Indian Fmg. 17(2):31.
1967.

514 INGRAM, W. R.
The lepidopterous stalk borers, associated with Gramineae in Uganda. Bull. Ent. Res. 49(2):367-383. London,
1958.

515 INGRAM, W. R., DAVIES, J. C. and McNUTT, D. N.
Agricultural Pest Handbook. Dept. of Agric. Uganda,
1966.

516 IRUTHAYARAJ, M. R. and RAJARATHNAM, S. A.
A note on the response of ragi (Eleusine coracana Gaertn.)
to calcium ammonium nitrate. Madras Agr. J. 1965, 52:
No. 1: 33-34.

517 ISAAC, P. V.
Report of the Imperial Entomologist (1944-45). Sci. Rep.
Indian Agric. Res. Inst. 1944-1945: 73-79.

518 ISH'KAWA, S.
Physiological studies on seed (Setaria italica) germination.
3. Germination inhibition by 8-hydroxy-quinoline and its
prevention with metal ions (preliminary report). Kumamoto J. Sci. Ser. B Sec. 2. Biol. 1960, 5: No. 1:5.

519 ITALIA AGRICOLA
Varieta di miglio ibrido foraggera per le zone aride. Ital.
Agric. 1959, 96, No. 11, 965-66.

520 ITO, S.
Mycological notes from Eastern Asia. II. Trans. Sap-
poro Nat. Hist. Soc. 14:87-96. 1935. (Cf. Rev. Appl.
Mycol. 15:829).

521 ITO, S. and KURIBAYASHI
The ascigerous forms of some graminicolous species of
Helminthosporium in Japan. J. Fac. Agric. Hokkaido
Univ. 29:85-125. 1931.

522 IVAKHNENKO, A.N.
Interrelations between the mycelia of Sphacelotheca panici-
miliacei Pers. Bub. and millet tissues in connection with
inheritance of smut resistance (from "Voprosy immuniteta
i ozdorovleniya Rastenii") Edited by Rodigin, M.N. Tr.
Khar'kovsk. Sel'skokhoz. Inst. 43. Kiev. Izdatel'stvo
Urozhad, 1964.

523 IVANOFF, S.S.
Inoculation tests with Phytomonas stewartii and Ph. vas-
culora. Phytopathology 25:21. 1935.

524 IVANOV, K.
An early-maturing millet. Zemledelie 5:63. 1967.

525 IYENGAR, C.V.K.
Some aspects of intraseasonal growth-variation in plants.
3. angiosperms: reproductive parts. Proc. Nat. Inst.
Sci. India (B) 1961, 27: No. 6:329-342.

526 IYER, A.K.Y.
Field crops of India with special reference to Mysore.
pp. 653. Supt. Govt. Press Bangalore 1947

527 JACQUES-FELIX, H.
Observations sur la variabilite morphologique de Coix
lachryma-jobi (15ᵉ note sur les graminees d'Afrique Tropicale). J. Agric. Trop. Bot. Appl. 1961. 8: No. 1-3:
44-56.

528 JACQUINOT, L. and CHEVASSUS-AGNES, A.
Preliminary tests for the measurement of the allogamy
rate by radioactive tracer technique on millet, P. typhoides Hubb. et Stapf. Ann. Centre Rech. Agron. Bambey 1960-1961. No. 20: 137-139.

529 JACQUOT, R., RAULIN, J., ADRIAN, J. and RERAT, A.
Composicion y valor alimenticio de las lagrimas de Job
(Lagrimas de San Pedro o trigo Adlay) Coix lachryma-
jobi). Arch. Venezol. Nutricion 1955, 6:3-21.

530 JAIN, T.C.
Soil moisture and evapotranspiration from these important
crop species (Pennisetum typhoides, Phaseolus acontifolius,
Ricinis communis) of arid zones. Agriculture (London)
15(3):133-137. 1967.

531 JAKIMENKO, A.F.
Maturation and harvesting of millet. Selek. Semenovodstvo 1952. No. 10:55-60. 1952.

532 JAKUSEVSKII, E. S.
Varieties of millet raised by the Kuban Experimental Station of the Institute of Plant Industry. Soviet Plant Industry Record 1940, No. 2. 105-7.

533 JAKUSEVSKII, E. S.
Japanese millet as a new fodder and industrial crop. Vestnik Socialisticeskogo Rastenievodstva (Soviet Plant Industry Record) 1940, No. 5. 188.

534 JAKUSEVSKII, E. S.
Setaria viridis. Kolkhoz. Proizv. 1950, No. 2. 26-27.

535 JANNONE, G.
Studi e ricerche di entomologia agraria in Etiopia e in Eritrea. I. Osservazioni su Tanymecus abyssinicus. Hust. (col. curcul.) condotte in Etiopia e in Eritrea dal 1939 al 1945. (Studies and researches on agricultural entomology in Abyssinia and Eritrea. I. Observations on T. abyssinicus in Abyssinia and Eritrea from 1939 to 1945.) Riv. Agric. Subtrop. Trop. 41(1-3): 43-54. Florence, 1947.

536 JANNONE, G.
Contributi alla conoscenze morfobiologica e sistematica dell'ortottero-fauna dell' Eritrea. VII. Studio sul comportamento della Schistocerca gregaria (Forsk.) in Eritrea dal 1939 al 1946 in rapporto di fattori ambientali e comparazioni col comportamento delle infestioni anofeliche. Bol. Lab. Ent. Agr. Portici 12:189-248. Portici, 1953.

537 JANSEN, G. R., DiMAIO, L. R., and HAUSE, N. L.
Amino acid composition and lysine supplementation of teff. J. Agr. Food Chem. 1962, 10:62-64.

538 JAROS, N. P.
Quantitative and qualitative composition of proteins and starch in grain of millets belonging to different ecological and geographical groups. Bull. Appl. Bot. Gen. Pl. Breed. 1965. 37: No. 1:50-58.

539 JAROSENKO, K. I.
Characteristics of growth and development in millets sown as a second crop. pp. 329-35. Abst. 6:55-136.

540 JASNY, N.
Competition among grains. Stanford U. Food Res. Inst. pp. 1-358. 1940.

541 JASOVSKIJ, I. V.
     An effective method of hybridization for millet.   Selek.
     Semenovodstvo 1960, No. 3:69-70.

542 JASTREBOV, F. S., KONSTANTINOV, S. I., MILAEV, Ja I.
     and STOLJARENKO, V. S.
     Emasculation of millet and sorghum by hot air.   Selek.
     Semenovodstvo 1964, 29: No. 3:71-72.

543 JAUHAR, P. O. and JOSHI, A. B.
     Cytological studies in some species of Panicum.   Cytolo-
     gia 1966. 31:153-159.

544 JAUHAR, P. O.
     Inter and intro-genomal chromosome pairing in an inter-
     specific hybrid, and its bearing on basic chromosome
     number in Pennisetum.   Genetics 1968. 39: 360-367.

545 JAUHAR, P. O. and JOSHI, A. B.
     New species of Panicum.   Bull. Bot. Surv. India. 1966.
     8: No. 1. 97-99 from Indian Sci. Abst. 1967: 3.   Abst.
     6425.

546 JAUHAR, P. O. and JOSHI, A. B.
     Accessory chromosomes in a new hexaploid species of
     Panicum.   Caryologia 1968. 105: 10.

547 JEN, H. -J., HSIN, S. -F. and LI, S. -W.
     Preliminary studies on the water requirements for germi-
     nation of different varietal groups of Setaria italica.
     Chung-kuo Nung-yeh Ko-hsueh 1963. No. 9:50-51.

548 JINKS, J. L. and HAYMAN, B. I.
     The analysis of diallel crosses.   Maize Genetics Newslet-
     ter 27:48-54. 1953.

549 JOCSON, A. T., GONZALES, T. T. and IBARBIA, R. A.
     Yield trials of millets in Lamao, Bataan, Philipp. J.
     Agric. 1959. 24:119-125.

550 JODPUR, P.
     Pusa Giant Napier - an Indian fodder grass.   Pl. Int.
     Rev. 2(3):24-25. 1965.

551 JOHNSON, D. T.
     Finger millet - Rhodesia's neglected crop (Eleusine cora-
     cana).   Rhod. Agric. J. 1968. 65: 87-91.   (Chilbero Agric.
     Coll., Rhodesia).

552 JOHNSON, J. C., LOWREY, R. S., MONSON, W. G. and BUR-
     TON, G. W.
     Influence of the dwarf characteristics on composition and
     feeding value of near isogenic pearl millets.   J. Dairy

Sci. 1968. 51: No. 9. 1423-1425. Bibl. 10. Anim. Sci.
Dept., Coastal Plain Exp. Sta. Tifton, Georgia.

553 JOHNSON, R.M. and RAYMOND, W.D.
The chemical composition of some tropical food plants.
1. Finger and bulrush millets (E. coracana and P. typhoides). Trop. Sci. 1964 6(1):6-11.

554 JOHNSTON, A.
Quarterly Report for April-June 1964 of the Plant Protec-
tion Committee for the South East Asia and Pacific Region.
FAO Publ., Bangkok, Thailand, 1964.

555 JONES, R.J.
Setaria progress report.   Pl. Introd. Rev. undated: 1:
No. 3: 59a-61a.

556 JORDAN, L.S., LYONS, J.M., ISON, W.H. and DAY, B.E.
Factors affecting performance of pre-emergence herbicides. Weed Sci. 16(4): 457-462. Illus. 1968.

557 JOSHI, A.B. and PATIL, B.D.
Napier-bajra hybrids.   Unpublished papers.   IARI.   1964.

558 JOSHI, B.S. and RACHIE, K.O.
Natural variability in Pennisetum typhoides S. and H.
Golden Jubilee Symposium of the Indian Agricultural Re-
search Institute, 1965.

559 JOSHI, L.M., RAYCHAUDHURI, S.R., BATRA, S.K., REN-
FRO, B.L. and GHOSH, A.
Preliminary investigations on a serious disease of Eleu-
sine coracana in the states of Mysore and Andhra Pradesh.
Indian Phytopathol. 19(3): 324-325.   1966.

560 JOSHI, M.G.
Radiation induced progressive nutrients in bajra (Pennisetum typhoides Stapf and Hubb). Curr. Sci. 1968:   37:
235-236.

561 JOSHI, M.S. and DASTANE, N.G.
Excess water tolerance of summer cereals.   Indian J.
Agron. 10:289-298. 1965.

562 JOSHI, S.D.
Report Preliminary Agric. Survey Almora District. Supt.
Govt. Press, Allahabad, U.P. Bull. 47. 1929.

563 JOSIFOVIC, M. and ARSENIJEVIC, M.
Prilog proucavanju helminthosporium monceras drechsl.
kao parazitr muhara i kukuruza (Contribution to the study
of H. monoceras as a parasite of millet and maize).   Zast.
Bilja, 15(79):283-292.   1964.

564 JOWETT, D.
     Use of rank correlation methods to determine food prefer-
     ences. Exp. Agr. 2(3):201-209. 1966.

565 JUJ, Ja. -P.
     Developmental phases and growth processes in Setaria
     italica. Sci. Rep. Sch. Biol. Sci. 1962. No. 2:171-173.

566 JURJEV, V. and others
     The institute's breeders on new varieties. Kolkhoz.
     Proizv. 1955. No. 12:40-41 and 42.

567 KABANOV, P. G.
     Makers of new varieties. Selek. Semenovdstvo 1957.
     No. 6:24-35.

568 KADAM, B. S.
     Rep. Dep. Agric. Bombay, 1936-1937. 1937.

569 KADAM, B. S.
     Rep. Dep, Agric, Bombay 1937-1938 1938

570 KADAM, B. S. and PATEL, S. M.
     Studies in field plot technique with P. typhoideum. Em-
     pire Jour. Exp. Agric. 5:219-230. 1938.

571 KALIAPPAN, R. and RAJAGOPAL, A.
     Effect of salinity on the South Indian field crops - dura-
     tion and grain setting in ragi (Eleusine coracana). Mad-
     ras Agr. J. 1968. 55: 125-128. (Fact. Agron. , Agric.
     Coll. Res. Inst. Coimbatore, India.

572 KALMYKOVA, G. , NEDBAILO, E. P. and SHALAMOV, F. F.
     Fertilization of prozo millet on the N. Caucasian calear-
     eous chemozem soils. Khimiya Sel'khoz 1968. 6: No. 10,
     18 (Ru: Donskoi sel'skokhoz Inst. Persianovka, Rostoviobli,
     U. S. S. R.)

573 KANAKARAJ, David S.
     A new genus and three new species of aphids from India.
     Indian J. Ent. 20(3): 175-180. 1958.

574 KANDASWAMY, M. and SUNDARM, N. V.
     Studies on rust resistant tenai varieties in Madras State
     (pp. 79-80). Proc. Sixth Sci. Wkrs. Conf. , Coimbatore
     (Madras) on 7 August, 1956 (1960): pp. 100.

575 KARUNAKARA Shetty, B. and MARIAKULANDAI, A.
     Studies on the effect of graded doses of nitrogen on the
     yield potentials of popular ragi (Eleusine coracana Gaertn)
     strains of the Madras State. Madras Agr. J. 1964. 5:210-
     15.

576 KASAHARA, Y. and KINOSHITA, O.
Studies on the control of the barnyard grass in the paddy
field. Proc. Crop Sci. Soc. Japan 22(3-4):7-10. 1954.

577 KAWAHARA, Eiji and WARAMATSU, Toshi Kazu
Studies on the germination of barnyard grass seeds. 1.
On the germination of barnyard grass seeds sown in the
irrigated and non-irrigated soils. Proc. Crop Sci. Soc.
Japan 33(1):64-68. 1964.

578 KAZAKENIC, L. I. and CEBOTAEV, N. F.
African millet. Opyt. Agron., No. 3, 6-12. 1941.

579 KEESE
Can Panicum miliaceum be grown as a grain crop in Germany. Dtsch. Landw. Pr. 68, 176 and 183-184. 1941.

580 KELKAR, S. P.
Maximisation of production by exploitation of the phenomenon of heterosis in bajra (Pennisetum typhoideum) S. &
H. Ann. Agric. Res. Abst. post-grad Res. wk. 1960-
1965. Nagpur Agric. Coll. Mag. 1966. Spec. Res. No.
111-112 (Abst.).

581 KEMPANNA, C. and KAVALLAPPA, B. N.
Quantitative assessment for nutritive quality of Eleusine
coracana (finger millet or ragi). Mysore J. Agric. Sci.
1968. 2: 324-329.

582 KEMPANNA, C. and TIRUMALACHAR, D. K.
Studies on phenotypic and genotypic variation in ragi
(Eleusine coracana) Mysore J. Agric. Sci. 1968. 11: No.
1: 29-34 (Univ. Agric. Sci. Bangalore, India).

583 KEMPTON, J. H.
Waxy endosperm in Coix and sorghum. J. Hered. 12(9):
396-400. 1921.

584 KEMPTON, J. H.
Elongation of mesocotyls and internodes in Job's Tears,
Coix lachryma-jobi (I). J. Wash. Acad. Sci. 31(6):261-
263. 1941.

585 KENNAN, T. C. D.
Matopos Research Station. Rhod. Agric. J. 1956. 53:
26-40.

586 KENNEDY, J. S.
The behaviour of the desert locust (Schistocerca gregaria
(Forsk)) (Orthopt.) in an outbreak centre. Trans. R. Ent.
Soc. Lond. 89(10):385-542. London, 1939.

587 KENNEDY-O'BYRNE, J.
Notes on African grasses; XXIX. A new species of Eleu-
sine from tropical and South Africa. Kew Bull. 1957.
No. 1:65-74.

588 KENNETH, R.
Studies on downy mildew diseases caused by Sclerospora
graminicola (Sacc.) Shroct. and S. sorghi Weston and
Uppal. In: Studies in Agricultural Entomology and Plant
Pathology, the Manges Press, Hebrew University: Jerusa-
lem, Israel. Scripta Hierosolymitana Publ. Hebrew Univ.
Jerusalem 18:143-170. 1966.

589 KENYA DEPT. AGRIC.
Annual Report for 1953. Vol. 2, Record of Investigations
(Item 6, 11). Nairobi, Kenya, 155. pp. 202-205.

590 KERLE, W.D.
Fodder millet. An ideal crop for quick spring and summer grazing. Agric. Gaz. N.S.W. 63: 301-5. 1942.

591 KEVAN, D.K. McE.
A study of the genus Chrotogonus audinot conville, 1000
(Orthoptera-acrididae). III. A review of available infor-
mation on its economic importance, biology, etc. Indian
J. Ent. 16(2):145-172. N. Delhi, 1954.

592 KHAN, A.M.
Productivity span of bajra 1 Napier grass hybrid. W.
Pakistan J. Agric. Res. 1966, 4, No. 1-2, 112-115
(Bibl. 3; Agric. Res. Inst. Tandojaur, E. Pakistan).

593 KHAN, M.D. and KHAN, A.A.
Cytological studies in bajra-napier grass hybrid. Proc.
12th Pakistan Sci. Conf., Hyderabad, 1960: Part III,
Abstracts: al-H18.

594 KHAN, Musahib-ud-Din and RAHMAN, Habib-ul-
Some cytogenetic studies in the inter-specific cross of
Pennisetum purpureum Schumach. and Pennisetum ty-
phoides Stapf and Hubb. Proc. 14th Pakistan Sci. Conf.
Peshawar, 1962: Part III, Abstracts B3-B4.

595 KHANNA, M.L.
Composition of bajra (Pennisetum typhoides) grain as influenced by fertilization. Indian J. Agron. 11:247-249.
1964.

596 KHASANOV, Yu. U., KOGAI, E.S. and KADYROVA, M.K.
Izchenie stabilnosti aldrina v nekotorykh ob'ektakh vneshni
sredy pri primenenii ego protiv gnusa. (A study of the
stability of the insecticide aldrin in some objects of the
environment when used to control blood sucking flies).

Gig. Sanit. 31(8):115-117. 1966.

597 KIHLBERG, R. and ERICSON, L. E.
     Amino acid composition and supplementation of teff. Nu-
     tritio et Dieta 1964, 6:151-155.

598 KILLINGER, G. B. and HARRIS, H. C.
     Pasture grass and legume responses to various fertilizer
     and management practices. Rep. Fla. Agric. Exp. Sta.
     1960:46-47.

599 KINRA, K. L. , NIJHAWAN, H. L. and RAO, S. B. P.
     Foliar spraying with urea boosts bajra yields.   Indian Fmg.
     16(7):15-54. 1966.

600 KIRILLOV, Yu. I.
     Floral biology of pearl millet.   Vestn. Sel-kh. Nauk.
     1963, No. 11, 42-48.

601 KIRILLOV, Yu. I.
     Aims and methods in breeding pearl millet.   Selek. Seme-
     novodstvo 1965, 30: No. 4:53-55.

602 KIRILLOV, Yu. J. and KOZULJA, I. E.
     The biology of pearl millet during the germination and
     shooting phases in relation to resistance to cold and salin-
     ity.   Rep. Agric. Sci. 1963, No. 10, 10-25.

603 KIRTIKAR, K. R. , BASU, B. D. and I. C. S.
     "Indian Medicinal Plants," Mohan Basu, Allahabad, 1933.

604 KISIELEV, D. A.
     A new variety of millet.   Selek. Semenovdstvo 1937: No.
     6:22-23.

605 KISSELBACH, T. A. and ANDERSON, A. A.
     Annual Forage Crops. Nebr. Agr. Exp. Sta. Bull. 206.
     1925.

606 KLEINIG, C. R. and NOBLE, J. C.
     (Commonw. Sci. & Indus. Res. Organ. , Div. Plant Ind.
     Deniliquin, N. S. W. Australia)
     1. The influence of weed density and nutrient supply in
     the field.   Aust. J. Exp. Agr. Anim. Husb. 8(32): 358-
     363. Illus. 1968.

607 KLIMOUSKII, D. N. and STASHKO
     Diastatic properties of millet and millet malt.   Spirto-
     Vodochnaya Prom. 17, No. 10/11, 7-9.   1940.

608 KNAPP, R.
     Effect of various temperatures on the germination of trop-
     ical and sub-tropical plants.   Angew. Bot. 1966, 39: No.
     6:230-241.

609 KNAPP, R.
     Effect of leaf litter from some African grasses and dry
     woodland species on germination. Ber. Dt. Bot. Ges.
     79(7): 329-335. 1966.

610 KOCK, G.
     Vinegar, a seed grain disinfectant. Fortschr. Landw. 7:
     226-227 (cf. Rev. Appl. Mycol. 11:566.) 1932.

611 KOCK, L.
     Djalibras. Korte Ber. 20. 7. 1919.

612 KOLESNIK, I. D.
     The cultivation of proso millet.   Moscow, All-Union Acad.
     of Agr. Sci., pp. 69. 1941.

613 KONOVALOV, I.
     On the question of the different correlations between lime
     and magnesia in the nutritive solution (theory of O. Loew).
     Zh. Opyt. Agron. 8(3): 257-280. 1907. USDA Exp. Sta.
     Rec. 19: 827.

614 KONSTANTINOV, N. N., ZHEBRAK F. A. and NIKOLISKII,
     IU K.
     Photo-periodism of millet and barley diploids and tetra-
     ploids. Dokl Akad. Nauk SSSR 1967(6): 1398-1400. 1966.

615 KOPP, A., and de CHARMOY, D'emmerez
     New observations of mosaic of sugar cane and streak of
     maize. Bull. Sta. Agron. Reunion Travaux Techniques
     3:1-10. 1932.

616 KORCENJUK, Ja T.
     The millet Veselyj-podol 38.   Selek. Semenovedstvo 1960.
     No. 4:71-72.

617 KORNICKE, F.
     Taxonomy. C. F. Hackel 1887 and Werth 1937. 1885.

618 KOROHODA, J. and JESMANOWICZ, A.
     Application of the graphical method of comparison of millet
     of different origin. Biul. Inst. Hodowl. Aklimatyz. Ros-
     lin 1957, No. 17:102-11.

619 KOSAREV, N. G.
     The new millet variety Kamysin 123.   Selek. Semenovodst-
     vo 1959, No. 2:72-73.

620 KOTTUR, G. L.
     Classification and distribution of cultivated plants of South
     Asia. Millets. Bull. Dep. Agric. Bombay 92. 1929.

621 KOUL, A. K.
Heterochromatin and non-homologous chromosome associations in Coix aquatica. Chromosoma 1964, 15:243-245.

622 KOUL, A. K.
Nucleolus in the genus Coix. Curr. Sci. 1965, 34:590-92.

623 KOUL, A. K.
Interspecific hybridization in Coix. I. Morphological and
cytological studies of the hybrids of a new form of Coix
with 2 n = 32 x Coix aquatica Roxb. Genetica 1966, 36
(3):315-324.

624 KOUL, A. K. and PALIWAL, R. L.
Morphology and cytology of a new species of Coix with 32
chromosomes. Cytologia 1964, 29: 375-386.

625 KOYANAGI, T., TAKANOHASHI, T. and HAREYAMA, S.
Effect on the cystine content in the hair, dark adaptation,
and urinary excretion of nitrogen, phosphorus and sulfur
when vitamin or milk is administered to children. J.
Jap. Soc. Food Nutr. 1964, 17:263-267.

626 KOZARENKO, M.
Nekotorye rezultaty vegetativnoi gibridizatsii zlakov.
(Some results of vegetative hybridization of cereals.)
Zemledelie 1954 (2): 114-115. Referat Zh. Biol., 1956.
No. 13910.

627 KRASSOWSKA, W. and TRZEBINSKI, J.
Treatment of millet for smut. Pam. Panst. Inst. Nauk.
Gosp. Wiejsk. Pulawach 1A(2):211-212. 1921.

628 KRASSOWSKA, W. and TRZEBINSKI, J.
Influence of disinfection of millet seed on the appearance
of Ustilago panici-miliacei. Pam. Panst. Inst. Nauk.
Gosp. Wiejsk. Pulawach 5A:273-276. 1924.

629 KRISHNAMURTHY, I. V. G.
Experiments on emasculation by hot water method in cumbu. Madras Agr. J. 37:133. 1950.

630 KRISHNAMURTHY, K.
Observations on the relation of tillering to nitrogen levels
in finger millet (Eleusine coracana Gaertn). Mysore J.
Agric. Sci. 1(1):37-43. 1967.

631 KRISHNAMURTHY, K.
Nutritive content of ragi varieties in relation to fertilizer
levels. J. Nutr. Diet. 1958. 5: 10-12 (Agronomy Div.
Agric. Coll., Univ. Agric. Sci. Bangalore - 24).

632 KRISHNAMURTI, B.
Some simple means of keeping food grains in storage free
from insect infestation and damage. Mysore Agric. J.
22(2): 40-45. Bangalore, 1943.

633 KRISHNASWAMI, N. and AYYANGAR, G. N. R.
Chromosome numbers in some Setaria species. Curr.
Sci. 1935, 3:559-560.

634 KRISHNASWAMI, N. and AYYANGAR
A note on the chromosome numbers of some Eleusine species. Curr. Sci. 1935, 4:104.

635 KRISHNASWAMI, N. and AYYANGAR, G. N. R.
Anatomical studies in the leaves of millets. J. Indian
Bot. Soc. 1942, 21:249-262.

636 KRISHNASWAMY, N.
Geography and history of millets. A translation from German. Curr. Sci. 6:355-358. 1938.

637 KRISHNASWAMY, N. and RAMAN, V. S.
A note on the number of some economic plants
of India. Curr. Sci. 1:376-378. 1949.

638 KRISHNASWAMY, N. and RAMAN, V. S.
Cytogenetics. Madras Agr. J. 43:239-242. 1956.

639 KRISHNASWAMY, N., SHETTY, B. V. and CHANDRASEKHA-
RAN, P.
Chromosome numbers of some Indian economic plants.
Curr. Sci. 23:64:65. 1954.

640 KRYSTALEVA, M. S.
The All-Union Show of new cultivars. Selek. Semenovodstvo No. 3:67-71. 1966.

641 KUBLAN, A.
Prospects and possibilities of millet cultivation - results
of breeding operations. Forschungsdienst 1943, 16:276-
83.

642 KUBLAN, A.
The reproduction of millet seed. Mitt. Landw. 58: 348-
9. 1943.

643 KUHNEL, Waltraude
Prufung verschiedener methoden zur infektion der rispe-
nune kolbenhirse (Panicum miliaceum L. und Setaria italica
L.) mit den brandpilzen Sphacelotheca panici-miliacei
(Pers.) Subak und Ustilago crameri Korn. Sowie untersuchung der dauer des infektions fahigen stadiums der
wirtspflanze. (Testing various methods for infecting com-

mon and Italian millet with smut fungi S. destruens and
U. crameri as well as investigations of the duration of
the infectible state of the host plant.) Nachrbl. Dtsch.
Pflschdienst N. F. 15(12):241-245. 1961.

644 KUHNAL, W.
Beobachtungen uber die abnahme der lebensdauer von
chlamydosporen verschiedener brandarten im boden (Observations on the decline in viability of chlamydospores of
various smut species in the soil). Zbl. Bakt., Abt. 2,
117(2):180-188. 1963.

645 KULKARNI, G.S.
Observations on the downy mildew of bajri and jowar.
Mem. Dep. Agric. India Bot. Ser. 5:268-273. 1913.

646 KULKARNI, G.S.
Smut (Ustilago paradoxa Syd. and Butl.) on sawan (Panicum frumentaceum Roxb.) J. Indian Bot. Soc. 3:10-11.
1922.

647 KULKARNI, G.S.
The smut of nachani or ragi (Eleusine coracana). Ann.
Appl. Biol. 9:184-186. 1922.

648 KULKARNI, N.B. and PATE, M.K.
Study of the effect of nutrition and temperature on the size
of spores in Piricularia setariae Nisik. Indian Phytopathol. 9:31-38. 1956.

649 KUMAKOV, V.A., KUZ'MINA, K.M.
Some features of photosynthetic activity and crop structure
in proso millet and spring wheat plants differing in time
of maturity. Fiziol. Rast. 1968. 15: No. 1:41-46 (Bibl.
6. Ru.e; Inst. Sel'khoz, Yugo-Vostoka Savator, U.S.S.R.)

650 KUMAKOV, V.A.
Photosynthesis and breeding for earliness. Selek. Semenovodstvo 1968. No. 1: 14-18 (Russian).

651 KURIEN, P.P. and DORAISWAMY, T.R.
Nutritive value of refined ragi flour (E. coracana). II.
Effect of replacing cereal in a poor diet with whole or refined ragi flour on the nutritional status and metabolism
of nitrogen, calcium and phosphorus in children (boys).
(Cent. Food Technol. Res. Inst. Mysore, Bangalore, India). J. Nutr. Diet. 4:2:102-109. 1967.

652 KURIEN, P.P.
Nutritive value of refined ragi (Eleusine coracana) flour.
I. The effect of feeding poor diets based on whole, refined and composite ragi flours on the growth and availability of calcium in albino rats. (Cent. Food Technol.

Res. Inst. Mysore, Bangalore, India). J. Nutr. Diet.
4(2): 96-101. 1967.

653 KURUP, C. K. R., VENKATRAMAN, K. V., TEJWANI, G. K.
and SASTRY, A. S.
Fertilizer trials with cumbu (P. typhoides) and cholan
(Andropogen sorghum) in rotation with cigar tobacco.
Indian J. Agron. 3:237. 1959.

654 KUTSCHERA, Lore
Wurzelatlas mitteleuropaischer ackerunkrauter und kultur-
pflanzen (Root Atlas of European Crops). Frankfurt am
Main: Dlg. Verlag. pp. 574. 1960.

655 KUWAYAMA, S.
Report on the distribution of and the conditions of injuries
by insect pests of important agricultural crops in Manchukou. Sangyobu Shiryo No. 33, 112 pp. Shinkyo Manchuria, Bur. Ind. Govt. Manchukou, 1938.

656 KUZINA, V. P.
Effect of simazine, atrazine and propazine on weeds and
yield of proso millet in the Steppe regions of Kuibyenhev
Province. Khimiya Sel'Khoz 1968. 6: No. 7: 54-5. (Ru.
Sel'skokoz Opyt. Stn. Bezen chuk Kuibyeshev Obl. U. S.
S. R.

657 KUZJMIN, V. P.
Selection of varieties for the Aknolinsk Steppe. Selek.
Semenovdstvo 4-5:25-34. 1946.

658 LA, A.
A forage variety of hybrid millet for arid zones. Ital.
Agric. 1959. 96: 965-966.

659 LABRUTO, Gaetano
On the chemical composition of the oil extracted from the
whole flour of "Eragrostis tef," a bread-cereal of eastern
Africa. Atti Soc. Pelorit. Sci. Fis. Mat. Nat. 9(1/2):
177-187. 1963.

660 LADD, E. F.
An active principle in the millet hay. Amer. Chem. J.
20(1898), No. 10. pp. 862-866. Exp. Sta. Rec. 10:794
(1898-99).

661 LAHIRI, A. N. and KHARABANDA, B. C.
Studies on plant-water relationships: effects of moisture
deficiency at various developmental stages of bulrush millet. Proc. Nat. Inst. Sci. India (B) 1965, 31, Nos. 1
and 2, 14-23.

662 LAHIRI, A. N. and KUMAR, V.
     Studies on plant-water relation. 3. Further studies on
     drought mediated alterations in the performance of bulrush
     millet. Proc. Nat. Inst. Sci. India (B) 32(3-4):116-129.
     1966.

663 LAL, B. M., JOHARI, R. P. and MEHTA, R. K.
     Some investigations on the oxalate status of Pusa giant
     napier grass and its parents. Curr. Sci. 1966, 35:125-
     126.

664 LAN, J. Z.
     Developmental morphology and histogenesis of reproductive
     structures of the millet, Panicum miliaceum L. 1. De-
     velopmental morphology of the inflorescence and flower.
     Acta Bot. Sin. 1958:7: No. 4:203-214.

665 LANE, C. B.
     Report of the dairy husbandman. New Jersey Sta. Rpt.
     1903, pp. 347-411. Exp. Sta. Rec. 16: 501.

666 LANGWORTHY, C. F. and HOLMES, A. D.
     Experiments in the determination of the digestibility of
     millets. USDA Bull. 525, 1917.

667 LANZA, M., REGLI, P. and BUSSON, F.
     Chemical study of the seed of Digitaria iburua Stapf.
     (Gramineae). Med. Trop. 1962, 22:471-476.

668 LEACH, R.
     Report of the mycologist. Rep. Dep. Agric. Nyasal.,
     1931. pp. 47-50. 1932.

669 LEAKEY, C. L. A.
     Dactuliophora, a new genus of Mycelia sterilia from
     Tropical Africa. Trans. Brit. Mycol. Soc., 47(3):
     341-350. 1964.

670 LEATHER, J. W.
     Water requirements of crops in India. Mem. Dep. Agric.
     India Chem. Ser. 1(8):133-184. 1910. Exp. Sta. Rec.
     23:332.

671 LeCLERQ, P.
     Surveys on the taste quality of sorghum and millet. Paper
     presented at the CCTA/FAO Symposium on Savannah Zone
     Cereals. Dakar 29 August - 4 Sept. 1962.

672 LEE, Y. N.
     A different kind of Setaria viridis. J. Jap. 40:370-371.

673 LEELA, R., DANIEL, V.A., RAO, S.V., HARIHARAN, K.,
    RAJALAKSHMI, D., SWAMIMATHAN, M. and PARPIA, H.A.B.
    Amino acid supplementation of proteins. 1. The effect of
    supplementing ragi (Eleusine coracana) and ragi diets with
    lysine, threonine and skim milk powder on the nutritive
    value of their proteins. J. Nutr. Diet. 1965 2:78-82.

674 LEFEVRE, P.
    Etude sur Busseola fusca Hmpsn. parasite du mais. Bull.
    Agric. Congo Belge 26(4):448-452. Brussels, Dec. 1936.

675 LELE, V.C. and DHANRAJ, K.S.
    A note on Helminthosporium rostratum Drechs. on ragi
    (Eleusine coracana (L) Gaertn). Curr. Sci. 25(17):446-
    447. 1966.

676 LEUCK, D.B. and BURTON, G.W.
    Pollination of pearl millet (Pennisetum typhoides) by in-
    sects. J. Econ. Entomol. 59(5):1308-1309. 1966.

677 LEUCK, D.B. et al.
    Fall army worm resistance in pearl millet. J. Econ.
    Entomol. 61 0 ᵖ 000-006 (1966)

678 LEVIN, A.M., ARZYBOV, N.A. and YURCHENKO, M.G.
    Effect of fertilizers on proso millet in Tambov Province.
    Khimiya Sel'khoz 1967. 5. No. 12. 21-2. (Bibl. 2; Ru:
    Plodov-ovoshich. Inst. Michurinsk, U.S.S.R.)

679 LEWICKI, S.
    The present state and possibilities of recovery of agricul-
    ture in Lubin County in view of the general Polish situa-
    tion. Bibljot. Pulaw. No. 21L 1-143. 1946.

680 LI, C.H., PAO, W.K. and LI, W.H.
    Interspecific crosses in Setaria. VI. Cytological studies
    of interspecific hybrids. J. Hered. 33:351-355. 1942.

681 LI, H.W., LI, C.H., and PAO, W.K.
    Cytological and genetical studies of the interspecific cross
    Setaria italica x S. viridis. Chin. J. Sci. Agr. 1944,
    1:229-248.

682 LI, H.W., MENG, C.J. and LIU, T.N.
    Field results in a millet breeding experiment. J. Amer.
    Soc. Agron. 1936, 28:1-15.

683 LI, Yao-Hui
    Effect of the chemical properties of calcaveous soils on
    the response of the yield of crops to phosphate fertilizer,
    Kuanchung district, Shensi. Acta Pedol. Sin. 13(1): 39-44.
    1965 (In Chin. with Eng. sum.).

63   Liang

684 LIANG, P.-Y., LEE (LI), Y.-L., and SHEN, L.-M.
Studies on Piricularia setariae Nishikado. Acta Phytopath. Sin. 1959: 5:89-99.

685 LINDSLY, J. B.
The composition, digestibility and feeding value of barnyard millet. (Panicum crus-galli). Mass. Sta. Rpt. 1900,
pp. 33-34. Exp. Sta. Rec. 13:377. 1901-02.

686 LIU, Tsung-shan
Resistance of millet varieties to different ecological-geo-
graphical populations of millet smut. Dokl. Mosk. Sel'-
khoz. Akad. K.A. Timiryazeva 31:136-143. 1957.

687 LLOYD, T.
Wild rice in Canada. Can. Geogr. J. 19(5):289-299.
1939.

688 LO, T.Y.
Nutritive value of the proteins of Chinese glutinous and
non-glutinous millets. Chin. J. Nutr. 1947, 2, 1-6.

689 LOBIK, V.I. and DAHLSTREM, A.F.
Improvement of methods for germination of wheat bunt
spores in the laboratory. Summ. Sci. Wk. Inst. Pl. Prot.
Leningrad, 1935. 172-178 (Cf. Rev. Appl. Mycol. 15:
786). 1936.

690 LOWIG, E. and EBNER, A.
Untersuchungen uber einige Weteigen Schaften des hirse-
kornes. Mitt. Inst. Pflanzenb. Pflanzenz. Landw.
Hochsch. Hohenheim 1944, No. 2: pp. 11.

691 LOZHNIKOVA, V.N.
The dynamics of natural gibberellins under conditions of
photo-periodic cycles. Dokl. Akad. Nauk SSSR 1966,
1968: No. 1:223-226.

692 LUCENKO, A.M., RUTTER, E.G., GRINSTEIN, Z.B., PEVZ-
NER, M.L. and SOBELEV, A.I.
Seed treatment with ultrasonics and yield. Vestnik Sel-kh.
Nauk. 1964, 9: No. 1: 128-130.

693 LUCERO, G.F.
Tests of herbicides for the control of weeds in lowland
rice fields. Philipp. Agric. 37(3):99-110. 1953.

694 LUNAN, M.
Mound cultivation in Ufipa, Tanganyika. East Afr. Agr.
J. 16:88-89. 1950.

695 LUTTRELL, E.S.
A Taxonomic revision of Helminthoseporium sativum and

related species. Amer. J. Bot. 42:85-125. 1955.

696 LUTTRELL, E. S.
      The identity, perfect stage and parasitism of Helmintho-
      sporium nodulosum. Phytopathol. 49:19. 1956.

697 LUTTRELL, E. S.
      Helminthosporium nodulosum and related species. Phyto-
      pathology 47:540-548. 1957.

698 LYMAN, C. M. et al.
      Essential amino acid content of farm feeds. J. Agr.
      Food Chem. 1956: 4: No. 12:1008-1013.

699 LYSAK, S. A. and BOGDANOVIC, M. V.
      Breeding millet for immunity to smut and other valuable
      economic characteristics. Selek. Semenovodstvo 1964,
      29: No. 6: 35-38.

700 LYSENKO, D. A.
      Wide spacing for millet. Zemledelie 1965, No. 5, 60-2.

701 LYSOV, V. N.
      Agrobiological classification of common millet (Panicum
      miliaceum L) Bull. Appl. Bot. Gen. Pl. Breed. 1952,
      29: No. 3:112-127.

702 LYSOV, V. N.
      Intraspecific relationships in the process of phasic devel-
      opment in common millet. Bull. Appl. Bot. Gen. Pl.
      Breed. 1962, 34: No. 3:93-105.

703 LYSOV, V. N.
      Crossing methods and evaluation of hybrid millet progeny.
      Bull. Appl. Bot. Gen. Pl. Breed. 1962, 34: number 3:
      106-114.

704 LYSOV, V. N.
      The longevity of Panicum miliaceum seeds. Tr. Prik.
      Bot. Genet. Selek. 38(1): 146-148. 1966.

705 LYSOV, V. N.
      Duration of a preserving of seed germinability in the proso
      millet Panicum miliaceum L. Tr. Prik. Bot. Genet.
      Selek. 38(1):146-148. 1966.

706 LYSOV, V. N. and ARTEM'EVA, N. N.
      Initial material for breeding millet for resistance to loose
      smut. Bull. Appl. Bot. Gen. Pl. Breed. 1962, 34: No.
      3:115-124.

707 LYSOV, V. N.
      Proso millet. Leningrad "Kolos" 1968, p. 224 (Bibl.

182: (Ru.) O. Horowitz.

708   LYUBENOV, Ya
        Results from certain herbicides tested in mixtures of
        maize and field beans. Rast. Nauki, Sofia 5(4): 103-117.
        Illus. 1968.

709   McALEESE, C. M.
        Weed control trials with auretryne.   Cane Growers Quart.
        Bull. 32(3):47. Illus. 1968.

710   McALPINE, D.
        The smuts of Australia. pp. 285. Melbourne 1910.

711   McCARRISON, R. and VISWANATH, B.
        The effect of manurial conditions on the nutritive and vita-
        min values of millet and wheat. Indian J. Med. Res. 14
        (2): 35-378. 1926.

712   McCARTY, M. K.
        Control of weeds in forage legumes and pasture. Res.
        Rep. N. Cent. Weed Contr. Conf. 1961 (n.d.) 116-117.

713   MACCHIATI, L.
        On the germination of old and mutilated seeds. Boll.
        Soc. Bot. Ital. 7-9:141-151. 1908.

714   McCLURE, F. J.
        Millet (cattail and foxtail) as the cariogenic component of ex-
        perimental rat diets. J. Dent. Res. 39(6): 1172-1176. 1960.

715   McCREARY, R. A., HOJJATI, S. M. and BEATY, E. R.
        Nitrates in annual forages as influenced by frequency and
        height of clipping. Agron J. 58(4): 381-382. 1966.

716   McEWEN, J.
        Raising cereal yields in Northern Ghana. Empire J. Exp.
        Agric. 1962. 30: No. 120: 330-334.

717   McGEE, C. R.
        Proso - a grain millet. Mich. Ag. Exten. Bul. 231.
        1941.

718   MACMILLAN, E. J.
        Agronomy and Seed Division. Annu. Rep. Dept. Agric.
        Orange River Colony 5:63-71. 1908-1909.

719   McRAE, W.
        Report on the Imperial mycologist. Sci. Rep. Agric.
        Inst. Pusa, 1921-22. 1922.

720 McRAE, W.
Rep. Bd. Sci. Adv. India. 1922-23. 1924.

721 McRAE, W.
Report of the Imperial mycologist. Sci. Rep. Agric. Res.
Inst. Pusa, 1928-1929. pp. 51-56. 1930.

722 McRAE, W.
Report of the Imperial Mycologist. Sci. Rep. Imp. Inst.
Agric. Res. Pusa, 1930-1931. pp. 73-86. 1932.

723 McRAE, W.
India: New diseases reported during 1928. Int. Bull. Pl.
Prot. 3:21-22, 1929.

724 MADRAS AGRIC....
Report on the operations of the Department of Agriculture
Madras Presidency for the year 1939-1940 (1941) pp. 81.

725 MADRAS AGRIC....
"Millets" (A. chapter). Mem. Dep. Agric. Madras Bot.
Ser. 30:1117-1119. 1954.

726 MADRAS AGRIC....
Short notes in the agenda, Department of Agriculture,
Madras. Proceedings, Conference of Workers on Millets,
Kohlapur, 1955. Indian Council of Agricultural Research,
New Delhi 1955.

727 MADRAS AGRIC....
Hybrid cumbu X. 3. Unpublished.

728 MADRAS AGRIC....
Co. 8 - a high yielding short duration ragi strain. Madras Agr. J. 1964, 51: p. 186.

729 MADRAS AGRIC....
H. B. 1 - the new hybrid cumbu seed doubles your yields.
Madras Agr. J. 1965, 52:460-461.

730 MADRAS AGRIC....
Co. 8 ragi. Madras Agr. J. 53:43. 1966.

731 MAGOON, M. L. and MANCHANDA, P. L.
A cytological study in some species of Paspalum. Indian
J. Genet Pl. Breed. 21(3):212-221. 1961.

732 MAHADEVAPPA, M.
Studies on heterosis in pearl millet (Pennisetum typhoides
Stapf and Hubb). Thesis, Madras Univ. 1965: from Indi-
an Sci. Abst. 1968:2. Abst. 7291.

733 MAHADEVAPPA, M.
Investigations on the inheritance of protein content in pearl
millet (Pennisetum typhoides Stapf and Hubb). Curr. Sci.
36(7): 186-188. 1967.

734 MAHADEVAPPA, M.
Investigations on seedling vigor in pearl millet (Pennise-
tum typhoides Stapf and Hubb). Proc. Indian Acad. Sci.
Sect. B 1967. 68:87-91. 1967.

735 MAHADEVAPPA, M.
Choice of type of cross and predicting hybrid performance
in pearl millet. (Pennisetum typhoides - Stapf & Hubb).
Madras Agr. J. 1967. 54:452-457.

736 MAHADEVAPPA, M.
Studies on heterosis in pearl millet (Pennisetum typhoides
S & H) 1. Expression of hybrid vigour and reciprocal
effects. Proc. Indian Acad. Sci.: Sect. B. 1968. 67: 180-
186. Univ. Agric. Sci. Bangalore, India.

737 MAHADEVAPPA, M.
Studies in heterosis in pearl millet Pennisetum typhoides
S & H) II. General and specific combining ability. Proc.
Indian Acad. Sci.; Sect. B. 1968. 67: 187-194. Univ.
Agric. Sci. Bangalore, India.

738 MAHADEVAPPA, M.
Diallel cross analysis of grain yield and some related
characters of pearl millet (Pennisetum typhoides S & H).
(Agric. Coll. & Res. Inst. Coimbatore, India). Mysore
J. Agric. Sci. 1968. 2(1): 49-51.

739 MAHADEVAPPA, M.
Studies on heterosis in pearl millet (Pennisetum typhoides
S & H) III. Genetic analysis of characters of peduncle
and primary ear. Proc. Indian Acad. Sci.; Sect. B.
1968. 68: 181-189.

740 MAHADEVAPPA, M.
Studies on heterosis in pearl millet (Pennisetum typhoides
S & H) IV. Genetic analysis of plant height, tillering,
yield of straw and yield of grain. Proc. Indian Acad.
Sci. Sect. B. 1968. 68: 210-220.

741 MAHADEVAPPA, M. and PONNAIYA, B.W.X.
Investigations on the formulation of selection index for
yield. Madras Agr. J. 1963, 50:84-85.

742 MAHADEVAPPA, M. and PONNAIYA, B.W.X.
A discriminant function for selection for yield in Eleusine
coracana Gaertn. Madras Agr. J. 1965, 52:47-51.

743 MAHADEVAPPA, M. and PONNAIYA, B.W.X.
A diallel cross study of some quantitative characters in pearl millet (Pennisetum typhoides S & H). Madras Agr. J. 1966, 53, 398-409.

744 MAHADEVAPPA, M. & PONNAIYA, B.W.X.
A note on utilizing male-sterile lines in single crosses of pearl millet. (Pennisetum typhoides S & H). Madras Agr J. 1966. 53, 510-513.

745 MAHADEVAPPA, M. and PONNAIYA, B.W.X.
A note on chromosome behaviour in some inbreds and hybrids of Pennisetum typhoides S & H. Madras Agr. J. 1967. 54, 85-88.

746 MAHADEVAPPA, M. and PONNAIYA, B.W.X.
Discriminant functions in the selection of pearl millet (Pennisetum typhoides Stapf and Hubb). Madras Agr. J. 54(5): 211-222, 1967.

747 MAHAPATRA, L.N.
Increasing production by interpolating tiller separated crop. Indian Fmg. 1965, 14: No. 10:20-22.

748 MAHAPATRA, L.N. and SINHA, S.K.
Clonal propagation of ragi (Eleusine coracana Gaertn). Sci. Cult. 1965. 31:92-93.

749 MAHAPATRA, L.N.
Eleusine coracana in Orissa. Indian J. Agron. 1967. 12 No. 1:19-24. (Bibl 3. Mixed Farm, Semiligudr. Koarput, Orissa, India.

750 MAHATA, D.N.
Rept. Econ. Bot. (Cotton). Rep. Dep. Agric. Cent. Prov. Berar. 1930-1931. 1931.

751 MAHUDESWARAN, K., NATARAJAN, M. and RAMACHAN-DRAN, A.
White ragi - a source for more protein. Madras Agr. J. 1966. 53:179-180.

752 MAHUDESWARAN, K.
Choice of position of leaf on main shoot to represent total leaf area and prediction of grain yield from plant characters in ragi; (Eleusine coracana Gaertn). Madras Agr. J. 1968. 55(9):380-386.

753 MAISURIAN, N.A.
Analysis of millet seed of Eastern Georgia, Caucasus. Bull. Polytech. Inst. Teflis 3:119-131. 1928.

754 MAKSIMOVA, E. V.
The effect of chlorides and sulfates on the respiration
rate and the activity of terminal oxidase in millet leaves.
Rostov-on-Don 241-24. 1962.

755 MAKSIMOVA, E. A. and MATUKHIN, G. R.
Effect of soil salinization on the respiration rate and terminal oxidase activity of millet leaves.   Fiziol. Rast.
12(3): 540-542. 1966.

756 MALICKI, L., PAWLOWSKI, F. and ZYZIK, A.
Yield and nutrient content of Japanese millet (Echinochloa
frumentacea) compared with maize (Zea mays spp. iden-
tata). Annls. Univ. Mariae Curie-Sklodowska Sect. E.
Agric. 1963 1964, 18:97-110.

757 MALLAMAIRE, A.
Acridiens migrateurs et acridiens sedentaires en Afrique
occidentale. Agron. Trop. 3(11-12): 630-634. Nogent-
sur-Marne, 1948.

758 MALM, N. R. and BECKETT, J. B.
Reactions of plants in the tribe Maydeae to Puccinia
sorghi Schw.   Crop Sci. 1962. 2:360-361.

759 MALUSA, K. V. and MIHAJLEC, V. I.
An evaluation of local millet varieties for their resistance
to diseases and insects.   Plant Prot. Insect. Dis. 1961,
No. 3:28-29.

760 MALYUGIN, E. A.
Methods of growing plants in sand semi-deserts of West
Kazakhstan.   Pustyni, U. S. S. R. i ikh Osvoenie 1954, No.
2:66-134.

761 MANCHANDA, P. L., UPADHYAY, M. K. and MURTY, B. R.
Evaluation of a world collection of genetic stocks of Pen-
nisetum.   2.   Pattern of genetic variation.   Indian J.
Genet. Pl. Breed. 1966, 26, 195-202.

762 MANDLOI, K. K. and TIWARI, P. K.
Kodon responds to nitrogen.   Indian Fmg. 16(1):24. 1966.

763 MANFREDI AGRICULTURAL EXPERIMENT STATION
Technical Report for the period 1 August, 1955 to 31 July,
1959.   Manfredi, Argentina, Agricultural Experiment Station, Pampean Reg. Centre: 1960. pp. 45.

764 MANITOBA DEPARTMENT OF AGRICULTURE
Annual Conference of Manitoba Agronomists, 1953. 1954,
pp. 63.

765 MANKIN, C.J.
     Seed treatment controls head smut of proso millet.   S.
     Dak. Fm. Home Res. 1959, 10: No. 4:18-19.

766 MANOHAR, M.S. and MATHUR, M.K.
     Germination studies of Pennisetum typhoides seeds treated
     with succinic acid under different water potentials.   Ann.
     Arid Zone 4(2):147-151. 1965.

767 MANOHAR, Man Singh and MATHUR, M.K.
     Effect of succinic acid treatment on the performance of
     pearl millet. (Pennisetum typhoides S & H).   Advancing
     Frontiers of Plant Sci. 17:143-147. 1966.

768 MARASSI, A.
     Eastern Peru and the Tingo Maria Agricultural Experiment
     Station.   Riv. Agric. Subtrop. Trop. 45:62-84 and 172-195.
     1951.

769 MARCHAL, A.
     The forms of Penicillaria cultivated in the Niger terri-
     tory. Agron. Trop. 1950. 5:582-592.

770 MARIAKULANDAI, A.
     Fodder scarcity and the possibilities of exploiting straw,
     processing in the Madras Presidency. Madras Agr. J.
     35:276-287. 1948.

771 MARIAKULANDAI, A. and MORACHAN, Y.B.
     Results of manurial trials in Madras State on millets.   1.
     Cholam and cumbu. Madras Agr. J. 53(4):163-170. 1966.

772 MARINICH, P.E., BREZHNEV, D.D. and PRUTSKOVA, M.G.
     (Eds)
     A guide for field inspection of crop cultivars   (Cereals
     and Millets). Moscow: 'Kolos,' 1966 p. 456 (Ru).

773 MARTIN, J.H. and LEONARD, W.H.
     Principles of field crop production.   pp. 519-531.   The
     Macmillan Company, New York, 1967.

774 MARTIN, W.J. and KERNKAMP, M.F.
     Variations in cultures of certain grass smuts.   Phytopath-
     ology 31:761-763. 1941.

775 MARUSEV, A.I. and IL'IN, V.A.
     The best millet cultivars in the South East.   Selek. Sem-
     enovodstvo 1965, 30: No. 5:7-9.

776 MASLOVSKII, A.D.
     Data on the immunity of millet to smut.   Agrobiologiya
     2:203-207. 1959.

777 MASSACHUSETTS HATCH STATION
     Fourth Annual Report 189 (pp. 14).   Exp. Sta. Rec. 3:
     699. 1891-1892.

778 MATHUR, R. L. , MATHUR, B. L. and BHATNAGAR, G. C.
     Blackening of bajra (Pennisetum typhoides Stapf and
     Hubb.) grains in earheads caused by Curvularia lunata
     (Wakk Boed. Syn. C. penniseti (Mitra) Boed. (Udaipur,
     Rajasthan).   Proc. Natl. Acad. Sci. India (B) 30(4):323-
     330. 1960.

779 MATHUR, R. L. , and MATHUR, B. N.
     Studies on the use of synthetic hormones against Striga
     lutea on bajra (Pennisetum typhoides) in Rajasthan.   Indi-
     an Phytopathol. 19(4):342-345.   1966.

780 MATHUR, S. B. , GOEL, L. B. and JOSHI, L. M.
     Fungi associated with seeds of Setaria italica and their
     control.   Proc. Int. Seed Test. Ass. 1967, 32, No. 3,
     633-638.   (Bibl. 3, E. d. f. , Div. Mycol. Pl. Path. Indian
     Agric. R & S. Inst. New Delhi).

781 MATSUDA, Kiyokatsu
     Investigations on the development of caryopsis in cereal
     plants. II.   Development of caryopsis in millet.   Proc.
     Crop Sci. Soc. Japan 13:150-155.   1941.

782 MATSUDA, Kiyokatsu
     Investigations on the development of caryopsis in cereal
     plants. III.   Development of caryopsis in Italian millet.
     Proc. Crop Sci. Soc. Japan 13:279-283.   1942.

783 MATZ, S. A.
     Chemistry and Technology of Cereals. (Ch. 8, pp. 177-
     189).   The AVI Publ. Co. , Westport, Conn.   pp. 732.
     1959.

784 MAZAEVA, M. M.
     Effect of magnesium fertilizers on plant development and
     yield.   Khimiya Sel'khoz: 1966. 4: No. 3: 166-172.
     (Bibl. 20. Russian).

785 MEDVEDEV, P. F.
     The first varieties of African millet.   Selek. Semenovod-
     stvo 14(10):53-58.   1947.

786 MEHRA, K. L.
     Differentiation of the cultivated and wild Eleusine species.
     Phyton 1963, 29: 189-198.

787 MEHROTRA, C. L. and GANGWAR, B. R.
     Studies on salt and alkali tolerance of some important ag-
     ricultural crops of Uttar Pradesh. J. Indian Soc. Soil

Sci. 12(2):75-83. 1964.

788 MEHTA, B.V. and SHAH, C.C.
Potassium status of soils of Western India. IV. Effect
of different levels of potassium in goradu soil on the uptake of cations by bajri (P. typhoideum). Indian J. Agric.
Soc. 82:473-490. 1958.

789 MEHTA, P.R. and CHAKRAVARTY, S.C.
A new disease of Eleusine coracana. Indian J. Agric.
Sci. 7:793-796. 1937 (RAM 17:454).

790 MELCHERS, L.E.
Studies on the control of the millet smut. Phytopathology
17:739-741. 1927.

791 MELCHERS, L.E. and JOHNSTON, C.O.
Sulphur and copper carbonate dusts as efficient fungicides
for the control of sorghum kernel smut and millet smut.
Phytopathology 17(1):52. 1927.

792 MELHUS, I.E. and VAN HALTERN, F.H.
Sclerospora on corn in America. Phytopathology 15:724-
730. 1925.

793 MELHUS, I.E. and VAN HALTERN, F.
Sclerospora graminicola on corn. Phytopathology 16(1):
85-86. 1926.

794 MELHUS, I.E., VAN HALTERN, F.H. and BLISS, D.E.
A study of Sclerospora graminicola (Sacc) Schroet. on
Setaria viridis (L) Beauv. and Zea Mays L. Res. Bull.
Ia. Agric. Exp. Sta. 111:297-340. 1928.

795 MENGESHA, M.H.
Eragrostis tef (Zucc) trotter, its embryo-sac development,
genetic variability and breeding behaviour. Diss. Abstr.
1965, 25: Order No. 65-2623:4338-4339.

796 MENGESHA, M.H. and GUARD, A.T.
Development of the embryo sac and embryo of teff, Eragrostis tef (Graminea) Can. J. Bot. 44(8):1071-1075.
1966.

797 MENGESHA, M.H.
Chemical composition of teff (Eragrostis tef) compared
with that of wheat barley and grain sorghum. Econ. Bot.
1966. 20:268-73.

798 MENGESHA, M.H., PICKETT, R.C. and DAVIS, R.L.
Genetic variability and inter-relationship of characters in
teff, Eragrostis tef (Zucc.) trotter. Crop Sci. 1965, 5:
No. 2:155-157.

799 MENON, P. Madhava
     Studies in cytoplasmic inheritance in Pennisetum typhoides
     S & H. Univ. of Madras, Ph.D. Thesis. 1958.

800 MENON, P. M. and DEVASAHAYAM, P.
     Studies on an interspecific hybrid in Pennisetum (P. squa-
     mulaum x P. typhoides x P. purpureum). Madras Agr.
     J. 1964, 51:p. 70.

801 MEREDITH, D.
     The Grasses and Pastures of South Africa. Cape Times
     Limited, Parow, C. P. 1955. pp. 362, 363, 524, 525.

802 MEREDITH, R. M.
     A review of fertilizer responses on the crops of Northern
     Nigeria. A paper presented at the FAO Conference on
     Soil Fertility and the Use of Fertilizers in West Africa,
     Ibadan, 1962.

803 METHOD, P.
     La production vegetale. Rapp. Minist. Agric. Prov.
     Queb. 1945:15-18.

804 MIHAJLEC, V. I.
     Local millet forms from L'vov Province and their im-
     provement by breeding. Symp. Sci. Wk. Sci. -Res. Inst.
     Agric. Anim. Husband. W. Distr. Ukrain. S. S. R. No.
     11:78-82. 1959.

805 MIHAJLEC, V. I.
     Local millet varieties from the Western districts of the
     Ukraine. Nauk. Prac. Nauk-Doslid. Inst. Zemlerob.
     Tvarynn. Zakhid Rai. URSR 1963:16:142-145.

806 MILES, J. F. and GRAY, S. G.
     Plant introduction notes No. 1 (pp. 7-11). Rep. Div. Pl.
     Ind. C. S. I. R. O. Aust. 1949. No. 5: pp. 45.

807 MILES, J. F. and GRAY, S. G.
     Trials of bulrush millet in Northern Australia. Rep. Div.
     Pl. Ind. C. S. I. R. O. Aust. No. 5: Plant Introduction
     Notes No. 1. 1949.

808 MILLER, P. E.
     Report of field crops work at the Morris Substation, 1917.
     Minn. Sta. Rep. Morris Substa., 1917. pp. 17-33.

809 MILLER, R. W. , HEMKEN, R. W. , VANDERSALL, J. H. ,
     WALDO, D. R. , OKAMOTO, M. and MOORE, L. A.
     Effect of feeding buffers to dairy cows grazing pearl mil-
     let or sudangrass. J. Dairy Sci. 48(10):1319-1323.
     1965.

810 MILLER, R.W., WALDO, D.R., OKAMOTO, M., HEMKEN, R.W., VANDERSALL, J.H. and CLARK, N.A.
Feeding potassium bicarbonate or magnesium carbonate to cows grazed on sudangrass or pearl millet. J. Dairy Sci. 46(6):621. 1963.

811 MILNE, D. and MOHAMMED, K.S. Ali
Handbook on field and garden crops of the Punjab. pp. 183. 1931.

812 MIMEUR, G.
Systematique specifique du genre Coix et systematique varietale de Coix lachryma-jobi. Morphologie de cette petite cereale et etude de sa plantule. Rev. Bot. Appl. 1951, 31:197-211.

813 MINOCHA, J.L., GILL, B.S. and SADHU SINGH
Desynapsis in pearl millet. J. Res. (India) 1968. 5: No. 2:32-36.

814 MIRCHANDANI, R.T.
Irrigated farming with particular reference to Sind. Indian Fmg. 8:501-503. 1947.

815 MISHRA, D. and MOHANTY, S.K.
A note on the response of crop seeds to pre-sowing treatment with B-nine. Trop. Agric. Trin., 43(4):347-349. 1966.

816 MISRA, A.P. and BOSE, A.
Varietal reaction of ragi to Helminthosporium nodulosum Berk and Curt. Indian Phytopathol. 19(3):310-311. 1966.

817 MISRA, D.K. and GOSWAMI, R.P.
Response of Pennisetum typhoides varieties to levels of nitrogen in Western Rajasthan. Indian J. Agron. 1967. 12: 116-120.

818 MISRA, D.K. et al.
Appraisal of two Pennisetum typhoides varieties in arid zone. Ann. Arid Zone 1966. 5: No. 1. 36-43; from Indian Sci. Abst. 1967. 3 Abst. 5978.

819 MISRA, D.K.
Bajra supplement. Indian Fmg. 1964. 14: No. 2:31-40.

820 MISRA, D.K. and BHATTACHARYA, B.B.
Agronomic studies on crops in arid zone. 4. Response of Pennisetum typhoides to micronutrients. Indian J. Agron. 11(2):178-184. 1966.

821 MISRA, D.K., BHATTACHARYA, B.B. and KUMAR, Vijay
Agronomic studies in arid zone. II. Germination as in-

fluenced by size of seed, temperature and time interval
on varieties of Pennisetum typhoides. Indian J. Agron.
11:264-266. 1966.

822 MISRA, D.K. and KUMAR, V.
Influence of depth of seeding on emergence, growth, crop
stand and yield of Pennisetum typhoides in arid zones.
Ann. Arid Zone 1964, 2:No. 2:1114-1122.

823 MISSISSIPPI AGRICULTURAL EXPERIMENT STATION
Grazing and pasture studies. Ann. Rep. Miss. Agric.
Exp. Sta. 72:47, 48-50, 51, 54, 64, 70, 75-76. 1958-
1959.

824 MITRA, M.
Morphology and parasitism of Acrothecium penniseti N.
Sp. (a new disease of Pennisetum typhoideum). Mem.
Dep. Agric. India Bot. Ser. 11:57-74. 1921.

825 MITRA, M.
A comparative study of species and strains of Helmin-
thosporium on certain Indian cultivated crops. Trans.
Brit. Mycol. Soc. 15:254-293. 1930.

826 MITRA, M.
Report of the Imperial Mycologist. Sci. Rep. Imp. Inst.
Agric. Res. Pusa, 1929-30. pp. 58-71. 1931.

827 MITRA, M. and MEHTA, P.R.
The effect of H-ion concentration on the growth of Helm-
inthosporium nodulosum and H leucostylum. Indian J.
Agric. Sci. 914-920. 1934a.

828 MITREVA, N. and PAVLOV, P.
Effect of ammonium nitrate and superphosphate upon millet germination. Selskostop. Nauka 1963, 2/8: No. 2:
250-251.

829 MIYAJI, Y. and KOKUBU, T.
Studies on the photoperiodic response in the Setaria italica.
3. On the flower formation under long day condition and
the reversal of the growth phase resulting from short
term treatment with short day in a late variety of Italian
millet. Proc. Crop. Sci. Soc. Japan 1957, 26: No. 2:
148-149.

830 MJAGKOV, N.V.
Methods of increasing the yield of millet. Selek. Semenovodstvo 1952. No. 2:25-31.

831 MONIZ, L. and BHIDE, V.P.
Root rot of cotton in Maharashtra and Gujarat States
caused by Macrophomina phaseoli (Maubl) Ashby var.

Indica N. Var. Indian Cott. Gr. Rev. 17(5):292-302.
1963.

832 MOOLANI, M. K.
Correlation of some plant characters with yield in T-55
bajra (Pennisetum typhoides (Pers). Indian J. Sci. Ind.
1968, 2, No. 1: 19-21. (Bibl. 10: Dep. Agron, Punjab
Agric. Univ. Hissar, India.)

833 MOOMAW, J. C. and KIM, D. S.
Selectivity of 2, 3, 5 trichino-4-pyridinos as a herbicide
for direct seeded flooded rice. Weed Res. 8(3) 1963-169.
Illus. 1968. Internat. Rice Res. Inst. Laguna, Philippines.

834 MOORE, A. W.
Nitrogen fixation in latosolic soil under grass. Pl. Soil
1963, 19:127-138.

835 MORTON, G. E., MAYNARD, E. J. and BRANDON, J. F.
Corn and hog millet for fattening lambs. Colo. Agric.
Exp. Sta. Bull. 71, 6 p. 1929.

836 MORTON, G. E., OSLAND, H. D. and BRANDON, J. F.
Fattening rations for hogs. Colo. Sta. Press. Bul. 81
(1932), pp. 15.

837 MOSCOW
Polyploidy in plants. Transactions of the conference on
plant polyploidy, 25-26 June, 1958. Tr. Mosk. Obshch.
Ispyt. Prir. 1962:5 pp. 376.

838 MOSCOW UNIVERSITY
Experimental morphogenesis. Moscow Univ. 1963 from
Referat. Zh. 1964.

839 MOUREAUX, C.
Modifications de la microflore d'un sol lateritique sous
differentes couvertures mortes (Modifications of the microflora of a lateritic soil under different mulches). VIth
Int. Congr. Soil Sci. Rep. C.407-412. 1956.

840 MOXON, A. L. and LARDY, H.
Manganese content of some South Dakota feeds. S. Dak.
Acad. Sci. Proc. 18, pp. 57-60. 1938.

841 MUDALIAR, V. T. S.
Common cultivated crops of South India. pp. 152-164.
Amudha Nilayam Pvt. Ltd., Madras, 1956.

842 MUHINA, V. A.
The diurnal course of photosynthesis and the translocation of assimilates during the photophase in some short-

day and long-day plants. Tr. Inst. Bot. V. L. Komarova
1960, Ser. 4: No. 14:167-187.

843 MUMINOV, H.
Studying and choosing Setaria forms and varieties and
methods of cultivating them in Kara-Kalpakia. Trans. Jn.
Sci. Sci. -Res. Inst. Univ. Min. Agric. Uzbec SSR 1962,
No. 1:99-105.

844 MUNDAY, H. G.
Possible rotation crops for Southern Rhodesia.   Rhod.
Agric. J. 8(1):59-68.   1910.

845 MUNDKUR, B. B.
Perfect stage of Sclerotium rolfsii in pure culture.   Indian
J. Agric. Sci. 4:779-781. 1934.

846 MUNDKUR, B. B.
A contribution towards a knowledge of Indian Ustilaginales.
Trans. Brit. Mycol. Soc. 23:105.   1939.

847 MUNDKUR, B. B.
A contribution to our knowledge of Indian Ustilaginales.
Trans. Brit. Mycol. Soc. 24:325, 1940.

848 MUNDKUR, B. B. and THIRUMLACHAR, M. J.
Revisions and Additions to Indian Fungi.   Mycol. Pap. 16.
Commonwealth Mycol. Inst. Kew. 1946.

849 MUNDKUR, B. B. and THIRUMLACHAR
Ustilaginales of India.   Commw. Mycol. Inst. Kew. 1952.

850 MURAKAMI, M.
Studies on the improvement by means of breeding of the
genus Coix. V. Genetical segregation in F_2 hybrids between C. ma-yuen and C. lachryma-jobi. Sci. Rep. Kyoto Prefect. Univ., Agric. 1961, 13:1-9.

851 MURAKAMI, M. and HARADA, K.
Studies on the improvement by means of breeding of the
genus Coix. I. On F_1 plants from the interspecific cross
C. ma-yuen x C. lachryma-jobi. Sci. Rep. Saikyo Univ.,
Agric. 1958, 10:111-120.

852 MURAKAMI, M. and HARADA, K.
Studies on the improvement by means of breeding of the
genus Coix. II. On tetraploid Coix raised by colchicine
treatment. Sci. Rep. Kyoto Prefect. Univ., Agric. 1959.
11:1-8.

853 MURAKAMI. M.
Studies on the improvement of breeding on the genus,
Coix. VII. On the sensitivity to x-rays of characters in

the F$_2$ hybrids of the progeny of the irradiated material. Sci. Rep. Kyoto Prefect. Univ., Agric. 1962, No. 14:1-11.

854 MURAKAMI, M., MIZUTIANI, T. and HARADA, K.
Studies on the improvement by means of breeding of the genus Coix. VI. On the resistance to submersion of some Coix plants. Sci. Rep. Kyoto Prefect. Univ., Agric. 1961, No. 13:10-22.

855 MURAKAMI, M., OYAGI, S. and HARADA, K.
Studies on the improvement by means of breeding of the genus Coix. VIII. The relation between number of cuts and green matter yield in C. ma-yeun, C. lachryma-jobi and some breeding strains, Sci. Rep. Kyoto Prefect. Univ., Agric. 1963, No. 15:1-11.

856 MURAKAMI, M., YASUDA, Y. and HARADA, K.
Studies on the improvement by means of breeding of the genus Coix. IV. Cytogenetical studies of C. ma-yuen, C. Lachryma-jobi and their F$_1$ hybrid. Sci. Rep. Kyoto Prefect. Univ., Agric. 1960, No. 12:11-17.

857 MURAKAMI, M., YODETAWA, U. and HARADA, K.
Studies on the improvement by means of breeding of the genus Coix. IX. Variation and heritabilities of some characters in F strains of C. ma-yuen x C. lachryma-jobi. Sci. Rep. Kyoto Prefect. Univ., Agric. 1964, No. 16:1-10.

858 MURTY, B. R.
Analysis of divergence in a world collection of sorghum and Pennisetum. Sols Afr. 11(1 and 2):453-462. 1966.

859 MURTY, B. R.
Response of hybrids of sorghum (jowar) and Pennisetum typhoides (bajra) to nitrogen. J. Postgrad. Sch. IARI, 1967: 5: No. 1:149-157. (Bot. Div. Indian Agric. Res. Inst. Delhi.)

860 MURTY, B. R., UPADHYAY, M. K. and MANCHANDA, P. L.
Classification and cataloguing of a world collection of genetic stocks of Pennisetum. Indian J. of Genet. Pl. Breed. (Catalogue of World Collection of Sorghum and Pennisetum) Vol. 27. Special Number. December, 1967.

861 MURZAMADIEVA, M. A.
The photosynthetic productivity and growth of millet in the dry steppe zone of Aktyubinsk Oblast. Vestnik Sel-Kh. Nauk. 10:9-14. 1963.

862 MURZAMADIEVA, M. A.
The water regime of millets in the arid steppe area of

Aktjubinsk Province Auyl Saruasyl. Gylymyn. Habarsysy
Vestn. Sel-Kh. Nauk. 1963. No. 8:35-43; from Referat.
Zh. 1964. Abst. 9. 55. 137.

863 MUSIENKO, V. F. and TIMCHENKO, V. A.
Paiza-tsennaya k'ormovaya kul'tura (Japanese millet -
available feed crop) Zhivotnovodstvo 1954(5):29-32. Referat. Zh. Biol., 1956, No. 76732.

864 MUSIL, A. F.
Seeds of grasses cultivated for forage or occurring inci-
dentally with crop seeds: The Genus Setaria. USDA
Leaflet (Mimeo), pp. 1-4. 1944.

865 MUSIL, A. F. and JACKSON, B. B.
Proceedings of the Association of Official Seed Analysts
Fortieth Annual Meeting, Washington, D.C. May 4-8,
1950, pp. 122.

866 MYSORE AGRIC. DEPT.
Annual Administration Report of the Agricultural Department, Bangalore for the year 1946-47. Bangalore, 1950.
pp. 402.

867 MYSORE DEPT. OF AGRIC.
Report from Agricultural Research Station, Bijapur (pp.
66-69). Mysore Agric. Cal. Yb. 1956-57. pp. 218.

868 MYSORE DEPT. OF AGRIC.
Brief account of the activities of the Agric. Res. Sta.,
Mandya (pp. 115-118). Mysore Agric. Cal. Yb. 1956-
57. pp. 218.

869 MYSORE DEPT. OF AGRIC.
Mysore Farmers are rushing to fertilize their dryland
ragi. Extension Wing and U.S. Agency International Development, Agricultural College, Hebbal, Mysore. Extension Leaflet, Sept. 1965.

870 NABOS, J.
Present state of research work on improved varieties and
cultivation methods of millet and sorghum in Niger. Sols
Afr. 11(1 and 2):365-374. 1966.

871 NABOS, J.
Brief report on millet (Pennisetum typhoides) and sorghum
in the Niger Republic. Sols Afr. 1967. 12: No. 2-3: 199-
203 (Fr. Stn. Rech. Agron., Tarna Maradi, Niger Republic.)

872 NAGAMATSU, T.
Genecological studies on Echinochloa crusgalli in rice
fields.   IV. On variation in the principal characters of E.
crusgalli from rice fields.   (Proc. Crop Sci. Soc. Japan
1952. 20:241-242.

873 NAGESWARA-RAO, P. and MURTHY, K. S.
Investigations into the mixed cropping in Mungari cotton
tract of Andhra Pradesh. Indian Cott. Gr. Rev. 19(3):
181-193. 1965.

874 NAIDU, B. A. and SINGH, D. J. C.
Studies on the preliminary trials with Dowpon.   Andhra
Agric. J. 7(L):8-9. 1960.

875 NAIDU, B. A. and VENKATESWARLU, B.
Influence of different levels of moisture supply on growth
and performance of Pennisetum typhoides S. & H. Andhra
Agric. J. 14(3):72-80. 1967.

876 NAIDU, B. A. and VENKATESWARLU, B.
Influence of soil moisture on sugar and nitrogen contents
of Pennisetum typhoides S. & H.   Andhra Agric, J, 1967
14, No. 5. 143-148. (Bibl. 13: Agric. Coll., Bapakala,
Andhra Pradesh, India.)

877 NAIR, M. K., RAMAN, V. S. and PONNAIYA, B. W. X.
Cytogenetics of certain derivatives of an interspecific
hybrid in Pennisetum. Madras Agric. J. 1964, 51:356-
357.

878 NAITHANI, S. P. and SISODIA, K. P.
Preliminary meiotic study in Pennisetum pedicellatum
Trin. Curr. Sci. 35; 343-344. 1966.

879 NAMBIAR, A. K., KRISHNASWAMY, P. and MENON, P. Madhava
Evolution of short duration hybrids in cumbu.   Proceedings 1st Scientific Workers Conference, Department of
Agriculture, Madras 1951. 1953.

880 NAMBIAR, A. K. and MENON, P. Madhava
Maximisation of production of cultivation of hybrid strains
with special reference to cumbu (pearl millet).   Madras
Agric. J. 38:95-100. 1951.

881 NAMBIAR, A. K. and MENON, P. Madhava
The exploitation of hybrid vigor in pearl millet.   Proceedings of the Conference of Workers on millets 14:76-85.
1955.

882 NANDA, G. S. and GUPTA, V. P.
General and specific combining ability in diverse types of

pearl millet. J. Res. (India) 1967: 4:343-347.

883 NARASIMHA RAO, D. V. and DAMODARAM, G.
Studies on correlation of certain plant characters to yield
in pearl millet (Pennisetum typhoides S. & H.) Andhra
Agric. J. 1964, 11:22-25.

884 NARASIMHA RAO, D. V. and PARDHASARATHI, A. V.
Studies on genetic variability in ragi. II. Phenotypic,
genotypic and environmental correlations between important
characters and their implication in selection. Madras
Agr. J. 1968. 55: 377-400.

885 NARASIMHA RAO, D. V. and PARDHASARADHI, A. V.
Studies on genetic variability in ragi. I. Phenotypic variation genetic advance and heritability of certain quantitative characters. Madras Agr. J. 1968. 55:392-397.

886 NARASIMHAN, M. J.
Report of the Department of Agriculture, Mysore. p. 21.
1921.

887 NARASIMHAN, M. J.
Report of the work done in the Mycological Section. Report of the Department of Agriculture, Mysore, 1921-1922.
1922.

888 NARASIMHAN, M. J.
Report of the work done in the Mycological Section during
1922-1923. Report of the Department of Agriculture, Mysore, 1922-1923, pt. 1, p. 4.

889 NARASIMHAN, M. J.
Ragi stored in underground pits. J. Mysore Agric. Exp.
Union 10:207. 1929.

890 NARASINHAM, M. J.
Administrative Report on Mycology, Mysore Department
of Agriculture, 1931-1932. pp. 32-35. 1933.

891 NARASINHAN, M. J.
Report of the work done in the Mycological Section. Report of the Department of Agriculture, Mysore, 1932-1933.
1934.

892 NARASIMHAN, M. J.
Early maturing ragi. Mysore Agric. Cal. Yb. 1940: 20-21.

893 NARASIMHAN, M. J. and THIRUMALACHAR, M. J.
Heteroecism in Uromyces setariae-italicae, the rust on
Italian millet. Mycologia 56(4):556-560. 1964.

894 NARASINGARAO, M.B.V., BALASUBRAMANIAN,C. and JAYA-
     RAMAN, M.V.
     Rainfall and crop yields in Madras State.   Madras Agr. J.
     37:475-483.   1950.

895 NARAYAN, K.N.
     Studies on the breeding behaviour of Pennisetum (P. clan-
     destinum) Curr. Sci. 23:269-270.   1954.

896 NARAYANA, Nuggihalli and NORRIS, Roland V.
     Studies in the protein of Indian foodstuffs.   Part I.   The
     proteins of ragi (Eleusine coracana).   Eleusinin the alco-
     hol-soluble protein.   J. Indian Inst. Sci. 11A(8):91-95.
     1928.

897 NARAYANASWAMY, S.
     The morphology of the bristles in Pennisetum and Setaria.
     Proc. 40th Indian Sci. Congr. Abstr. 37.   1953.

898 NARESH, Chandra
     Morphological studies in the Gramineae.   IV.   Embryology
     of Eleusine indica Gaertn and Dactyloctenium aegyptium
     (Desf.) Beauv. Proc. Indian Acad. Sci. Sect. B. (58(3):
     117-121.   1963.

899 NAZARENKO, K.S.
     On the farms with the best cultivars.   Selek. Semenovodst-
     vo No. 3:51-59.   1966.

900 NELSON, C.E. and ROBERTS, S.
     Proso grain millet as a "catch crop" or "second crop"
     under irrigation.   Sta. Circ. 376 Wash. Agric. Exp. Sta.
     1960, pp. 5.

901 NEMLIENKO, L.M.
     The millet Mironovka 85.   Selek. Semenovodstvo 1959.
     No. 6:64-65.

902 NENAROKOV, M.I.
     Utilization of water meadows in the southern forest/steppe
     and northern steppe.   Zemledelie 4(5):89-97.   1956.

903 NEVALENYI, V.Z.
     Yield of millet after repeated sowing during crop rotation.
     Tr. Ukrain. Nauch. -Issledovatel. Inst. Rastenievodstva,
     Selektsii i Genet.   5:59-62.   1959.

904 NEW SOUTH WALES. Dept. of Agriculture
     Annual Report 1967. p. 172.

905 NEWLANDER, J.A.
     The digestibility of artificially dried roughages.   Vermont
     Sta. Bul. 100, pp. 12.   1935.

906 NEZAMUDDIN, S. and MEHDI, Akhtar
     A.404, A.407: The ragi varieties for Bihar.   Indian Fmg.
     1966.   15: No. 11 p. 23.

907 NICOU, R.
     L'influence de quelques techniques de culture sur les ren-
     dements des mils et sorghos.   Colloque CCTA/FAO sur
     les cereales des zones des savane.   Dakar.   29 August -
     4 September, 1962.   Bureau des Publications, Watergate
     House, London, W.C. 2.

908 NIEHAUS, M. H. and PICKETT, R.C.
     Sorghum evaluation at the southern Indian forage farm in
     1962.   Res. Progr. Rep. Purdue Univ. 1963, No. 57:
     pp. 5.

909 NIGERIAN DEPT. OF AGRIC.
     Annual Report - Northern Region 1954-1955.   Part 2. Re-
     search and Specialist Services.   Kaduna (n. d.) April, 1956.
     (e) Rotations and Fertilizers, pp. 9-15.

910 NIKITINA, A. V.
     Effects of fertilizers and micro-elements on improvement
     of resistance of millet to Sphacelotheca panici-miliacei
     (Pers.) Bibl. Uchenye Zap. Kharkov. Univ. 1963. No.
     141:140-145; Referat Zh. 1964. Abst. 22. 55. 137.

911 NIQUEUX, M.
     Les millet.   Morocco, Direction de l'Agriculture, du Com-
     merce et des Forets.   Cah. Rech. Agron. 3. 1950, 401-
     479.

912 NIRODI, N.
     Studies on the Asiatic relatives of maize.   Ann. Mo. Bot.
     Gdn. 1955, 42:103-30.

913 NIRODI, N.
     Studies on some Asiatic relatives of maize.   Diss. Abstr.
     1954, 14: Publ. No. 8988: p. 1513.

914 NISHIHARA, Natusuki
     Eye-spot disease of pearl millet in Japan.   Nat. Inst.
     Anim. Ind. (Chiba) Bull. 11:23-29.   1966.

915 NISIKADO, Y. and MIYAKE, C.
     Studies on the Helminthosporium of the rice plant.   Ber.
     Ohara Inst. 2:133-194.   1922.

916 NISIKADO, Y.
     Studies on the rice blast fungus.   I. Ber. Ohara Inst. 1:
     171-217.   1917.

917 NISIKADO, Y. and MIYAKE, C.
On a new Helminthosporium parasitic on Panicum crusgalli. I. Ber. Ohara Inst. 2:597-612. 1925.

918 NISIKADO, Y.
Studies on the rice blast fungus. Jap. J. Bot. 3:239-244.
1927.

919 NISIKADO, Y.
Leaf blight of Eragrostis major Host caused by Ophiobolus
kusanoi N. sp., the ascigerous stage of Helminthosporium.
Jap. J. Bot. 4:99-112. 1928.

920 NODA, A. and HAYASHI, Z.
Studies on the coleorhiza of cereals. Proc. Crop Sci.
Soc. Japan 1959, 28: No. 1:17-19.

921 NOGGLE, J. C. and FRIED, M.
A kinetic analysis of phosphate absorption by excised
roots of millet, barley and alfalfa. Soil Sci. Soc. Amer.
Proc. 24:33-35. 1960.

922 NORMAN, M. J. T.
Grazing and feeding trials with beef cattle at Katherine,
N. T. (Australia). Tech. Pap. 12 Div. Ld. Res. Reg.
Surv. C. S. I. R. O. pp. 15, 1960.

923 NORMAN, M. J. T.
Performance of annual fodder crops under frequent defoli-
ation at Katherine, N. T. (Australia). Tech. Pap. 19
Div. Ld. Res. Reg. Surv. C. S. I. R. O. 1962, p. 11.

924 NORMAN, M. J. T. and WETSELLAR, R.
Performance of annual fodder crops at Katherine, N. T.
(Australia). Tech. Pap. 9 Div. Ld. Res. Reg. Surv.
C. S. I. R. O. 1960, pp. 16.

925 NORMAN, M. J. T., and BEGG, J. E.
Bulrush millet (Pennisetum typhoides Burm.) at Katherine, N. T. J. Aust. Inst. Agric. Sci. 34:2:59-68 (1968).

926 NORMAN, M. J. T., and PHILLIPS, L. J.
The effect of time of grazing on bulrush millet (Pennisetum typhoides). Aust. J. Exp. Agr. Anim. Husb. 8
(32):288-293. Illus. 1968.

927 NOVITSKAYA, Yu E.
The importance of presowing hardening of plants to
drought in solutions of certain microelements. Tr. Bot.
Inst. Akad. Nauk SSSR Ser. 4(12):74-94. 1958.

928 NUTTONSON, M. Y.
Ecological crop geography in China and its agro-climatic

analogues in North America. Washington, American Institute of Crop Ecology International Agro-Climatological Series Study No. 7. 1947. pp. 28.

929 NYASALAND AGRICULTURAL QUARTERLY JOURNAL
Sorghums and millets. Nyasaland Agric. Quart. J. 1951,
10: No. 3:80-81.

930 NYASALAND DEPARTMENT OF AGRICULTURE
Report for the Year 1950. Pt. II. Zomba, 1952, pp. 36
(5. includes millet trials p. 23). Zomba, pp. 36. 1952.

931 NYE, I. W. B.
The insect pests of Graminaceous crops in East Africa.
Report of a survey carried out between March 1956 and
April 1958. Colonial Research Studies No. 31. London,
Colonial Office, H. M. S. O., 1960.

932 NYE, P. H. and FOSTER, W. N. M.
The relative uptake of phosphorus by crops and natural
fallow from different parts of their root zone. J. Agric.
Sci. 1961, 56: No. 3: p. 99-306.

933 NYE, P. H. and HUTTON, R. G.
Some preliminary analyses of fallows and cover crops at
the West African Institute for Oil Palm Research, Benin.
J. W. Afr. Inst. Oil Palm Res. 2:237-243. 1957.

934 OELKE, E. A. and MORSE, M. D.
Control of barnyard grass in continuously flooded rice
fields. Rice J. 72:2: p. 16; 18-20 (1968).

935 OISHI, T.
Sideridis unipuncta. Haw. and its enemies. J. Plant
Prot. 27(7):488-490. Tokyo, 1940.

936 OLDENHOVE, H.
Some considerations concerning C. lachryma-jobi. C. R.
Semaine Agric. Yangambi, Part I. Publ. Inst. Nat. Agron.
Congo Belge 1947, Hors. Ser.: 374-379.

937 ONO, T.
Investigations on the production of polyploids of barley and
other cereals. Jap. J. Genet. 1947, 22:55-56.

938 ORLICZ, Anna
Szezatki reslinne z okresu rzymskiego z wykopalisk archeo-
logicz nych w Wasoszu gornym kolo klobucka (Plant remains of Roman times from archeological excavation at
Wasosz Gorny near Klobuck). Folia Quarternaria 27:1-9.
1967.

939 ORLOV, P.
Problems of breeding and seed production with millet.
Selek. Semenovodstvo 1937. No. 8-9: 49-51.

940 OSTROVSKAYA, L. K., OVCHARENKO, G. A., RASTORUEVA,
L. T., et al.
Uptake, movement and assimilation of iron in plants.
Agrokhimiya 1966, No. 1:100-108.

941 OSTROVSKII, N. I. [OSTROVSKY, N. I.]
Fecundity in females of Stenodiplosis panici Rodd. as influenced by feeding conditions at larval stage. C. R.
Acad. Sci. U. R. S. S. (n. s.) 52(6):555-556. Moscow,
1946.

942 OUBRECHT, J.
Technological procedures in the cultivation of teff - Beitr.
Trop. Subtrop. Landwirt. Tropenveterinarmed. 5 (I): 15-
21. 1967. (Ger., Fr., Russ. and Span. summ.).

943 OVERSEAS FOOD CORPORATION
Annual Report for year ending 31 March, 1955 (Tanganyika).

944 OXLEY, T. A.
Grain storage in East and Central Africa. Report of a
survey (Oct. 1948 to Jan. 1949). Colonial Research Publication No. 5. London, H. M. S. O., 1950.

945 OYENUGA, V. A.
The composition and nutritive value of certain feeding-
stuffs in Nigeria. 2. Concentrates. Empire J. Exp.
Agric. 23(91/2):171-187. 1955.

946 OZBEK, Nurinnisa
Factors affecting the amount of available soil phosphorus,
A-value. In: Proceedings of the Symposium on Isotopes
in Plant Nutrition and Physiology, 5-9 September, 1966,
Vienna, Austria, International Atomic Energy Agency:
Vienna, Austria. 35-46. 1967.

947 PALANIANDI, V. G. and ANANTHAPADMANABHAN, C. O.
Suitable ragi varieties for lower Bhavani project area in
the first crop season. Madras Agr. J. 55:333-334. 1966.

948 PANG, Shih Chuan, CHANG, Sian Tze and WU, Ching Fung
A preliminary study of increasing the salt tolerance of
soybean and Italian millet (Setaria italica). Acta Bot. Sin.
1964, 12: No. 1:64-74).

949 PANTULU, J.V.
Meiosis in an autotriploid pearl millet. Caryologia 1968.
21: No. 1: 11-15.

950 PANTULU, J.V. and VENKATESWARLU, J.
Morphology of the pachtene chromosomes of Pennisetum
purpursum Schumach. P.B.A. 1969. No. 664. p. 93.
Genetica 1968. 39: 41-44.

951 PARK, M.
Technical Report, Mycology Division, Ceylon Department
of Agriculture, 1929. 1930.

952 PARK, M.
Report on the work of the Mycological Division. Administrative Report of the Director of Agriculture, Ceylon 1931.
pp. 103-111. 1932.

953 PARSONS, D.J.
The systems of agriculture practised in Uganda. No. 1. Introduction and Teso systems; No. 2. The plantain-robusta
coffee systems with a note of the plantain-millet-cotton areas;
No. 3. The northern systems. Pt. 1. The Lango-Acholi system. Pt. 2. The West Nile system; No. 4. Montane systems;
No. 5. Pastoral systems. Mem. Res. Div. Dep. Agric.
Uganda Ser. 3: 1969. pp. 70(1):57 (No. 2):66 (No. 3): No. 4,
5 pp. 30 and 27.

954 PARTHASARATHY, S.V. and SUNDERARAJ, D.D.
Fodder grasses of Madras. Madras Agr. J. 35:379-383.
1948.

955 PATEL, B.M., SHAH, B.G. and MISTRY, V.V.
A study on the fodders of Hissar District in the Punjab.
Indian J. Agric. Sci. 28:597-606. 1958.

956 PATEL, B.M. and SHAH, B.G.
Studies on the composition of some cereal straws in Kaira
District. Indian J. Agric. Sci. 1959, 29:19-25.

957 PATEL, M.K. and THIRUMALACHAR, M.J.
Notes on some Xanthomonas species described from South
India. Curr. Sci. 34(14):436-437. 1965.

958 PATEL, Z.H.
A sugary mutant in pearl millet (Pennisetum typhoideum)
Proc. 28th Indian Sci. Congr. Benares 1941: Part III.
Sect. Agric. Abst. 40:p. 258.

959 PATIL, B.D.
Two new Pusa Napiers. Indian Fmg. 12(11):20, 23.
1963.

960 PATIL, B.D., HARDAS, M.W. and JOSHI, A.B.
Auto-alloploid nature of Pennisetum squamulatum Fresen.
Nature 1961, 189:419-420.

961 PATIL, B.D. and JOSHI, M.G.
Sterility in pearl millet (Pennisetum typhoides Stapf. &
Hubb.) Curr. Sci. 1962, 32: p. 38.

962 PATIL, B.D., SINGH, Amar
An interspecific cross in the genus Pennisetum involving
two basic numbers. Curr. Sci. 1964, 33:p. 255.

963 PATIL, B.D. and VOHRA, S.K.
Desynapsis in Pennisetum typhoides Stapf and Hubb. Unpublished (1960).

964 PATIL, P.C.
The crops of the Bombay Presidency, their geography and
statistics. Bull. Dep. Agric. Bombay 109. 1922.

965 PATIL, P.K. and PATIL, S.M.
Need for research on bajri in Bombay State. p. 156.

966 PATIL, P.L., KULKARNI, N.B. and MORE, B.B.
Curvularia leaf blight of bajra. Pennisetum typhoides S.
& H. in India. Mycopathol. Mycol. Appl. 28(4):348-352.
1966.

967 PATNAIK, M.C.
Timely sowing pushes up ragi yield in Orissa. Indian
Fmg. 1968. 18. No. 8:14-17.

968 PAUL, W.R.C.
Progress in pasture work in the humid lowland region (B.
ramasa). Trop. Agriculturist, 1948. 104:141-50.

969 PAVLOV, P.
Some biochemical changes in millet grown under various
light regimes. Rast. Nauki, Sofia, 1965, 2, No. 7, 43-
49.

970 PERDOMO, J.T., SHIRLEY, R.L. and ROBERTSON, W.K.
Soil and Crop Science Society of Florida. 26th Annual
Meeting, Clearwater, Florida, 6 - 8 December, 1966.
Proc. Soil Crop Sci. Soc. Fla. 1966. 26, p. 412.

971 PEREIRA, H.C., WOOD, R.A. and BRZOSTOWSKI, H.W.,
et al.
Water conservation by fallowing in semi-arid tropical East
Africa. Empire J. Exp. Agric. 1958, 26:213-228.

972 PEREVERZEVA, V.A.
The cultivation of foxtail millet in the Central Zone of

Stavropol'sk. S. -Kh. Inst. 1956(7):119-128. 1956.

973 PERKINS, H. F.
Short term procedure for correlation of soil test values
for phosphorus with growth of browntop millet (Panicum
ramosum). Agron. J. 57(4):410-411. 1965.

974 PERRIN, de Brichambaut G.
Especes introduites et spontanees du genre Pennisetum.
Morroco, Direction de l'Agriculture du Commerce et des
forets. Cah. Rech. Agron. 3. 1950, 369-400.

975 PETA, P. R.
A comparison of pearl millet and sudangrass as pastures
for lactating dairy cows with special emphasis on milk
fat. Diss. Abstr. (B) 1967. 27: 2934B-2935B.

976 PETERS, L. V.
Bulrush millet (pearl millet) Pennisetum typhoides in
Uganda and East Africa. Sols Afr. 1967. 12: 153-157.

977 PHELPS, C. W.
Fertilizer experiments on Hungarian grass. Rpt. Storrs
Agric. Exp. Sta. 1893, pp. 130-135. USDA Exp. Sta.
Rec. 6:405. 1894-1895.

978 PHILIPPINES DEPT. AGRIC.
What others say about Adlay. Philipp. Agric. Rev. 14(2):
1968-77. 1921.

979 PHILLIPS, L. J. and NORMAN, M. J. T.
Fodder crop-cash crop sequences at Katherine, N. T. (Aus-
tralia). Tech. Pap. 20 Div. Ld. Res. Reg. Surv. C. S. I.
R. O. pp. 12, 1962.

980 PHILLIPS, L.J. and NORMAN, M.J.T.
The effect of chisel ploughing and inter-row cultivation on
soil water content and yield of bulrush millet (Pennisetum
typhoides) at Katherine, N. T. Aust. J. Exp. Agric. Anim.
Husb. 1966, 6:48-55.

981 PICHNER, F.
Calculation of the quantities of dusts required for various
kinds of seeds. Nachr. Schadlbekampf. 13:17-18. 1938.

982 PIPER, C. V.
Forage crops and forage conditions in the Philippines.
Philipp. Agric. Rev. 4(8):394-428. 1911.

983 PISAREV, V. E.
Along the road of Micurin. Zemledelie 1955. 10:37-42.

984 PISHNAMAZOV, A. M.
Growing seeds of sudan grass and Japanese millet at after-
harvest sowing. Zemledelie 6. 85. 1959. Referat. Zh.
Biol., 1960, No. 25062.

985 PKHAKADZE
Cultivation of millet (chumizy) in Georgia. Tr. Gruzinsk.
Sel'skokhoz. Inst. 1955 (42-43):67-87. 1955. Referat.
Zh. Biol., 1957. No. 69253.

986 PLOTNIKOV, N. Ja
Varietal sowings of millet. Selek. Semenovodstvo 1946.
13: No. 6:62-65.

987 PLOTNIKOV, N. Ja
New millet varieties. Selek. Semenovodstvo 1947. 14:
number 2: p. 73.

988 POHL, R. W.
Notes on Setaria viridis and S. faberii (Gramineae); Brittonia 1962, 14: No. 2:210-213.

989 POKHRIYAL, S. C. and MANGATH, K. S.
Studies on the nature of protogyny for production of hybrid
seeds in pearl millet. LABDEV 34(3):192-194. 1965.

990 POKHRIYAL, S. C., MANGATH, K. S. and GANGAL, L. K.
Genetic variability and correlation studies in pearl millet
(Pennisetum typoides) (Burm. F.) Stapf and C. E. Hubb.
Indian J. Agric. Sci. 37:77-82. 1967.

991 POKHRIYAL, S. C., RAO, S. B. P., MANGATH, K. S. and
RAJPUT, D. D.
Effect of inbreeding on yield and other characters in pearl
millet. Indian J. of Genet. Pl. Breed. Vol. 26(2): 210-
216. July, 1966.

992 POKHRIYAL, S. C., PRIYAVRATHA RAO, S. B., and MANGATH,
K. S.
Pistil-less mutant in bajra (Pennisetum typhoides Stapf
& Hubb.). Sci. Cult. 33:192. 1967.

993 POKHRIYAL, S. C., MANGATH, K. S. and RAO, S. B. P.
Hybrid vigour in pearl millet (Pennisetum typhoides S. &
H.). Indian Agr. 1967: 11:55-61.

994 POPOV, G.
Some notes on injurious acrididae (Orthoptera) in the Su-
dan-Chad area. Ent. Mon. Mag. 95(1139):90-92. London,
1959.

995 POPOV, G. I.
Constructive selection. Selek. Semenovdstvo. 1940, No.

5: 9-11.

996 POPOV, G. I.
Breeding methods at the Kazanj State Breeding Station.
Selek. Semenovodstvo 1949. No. 11: 44-47.

997 PORTER, R. H. , YU, T. F. and CHEN, H. K.
Effect of seed disinfectants on smut and yield of millet.
Phytopathology 18:911-918. 1928.

998 PORTERES, R.
L'aire culturale du Digitaria iburua Stapf, cereale mineure
de L'ouest-Africain. Agron. Trop. 1(11-12):589-592.
1946.

999 PORTERES, R.
Les plantes indicatrices du niveau de fertilite du complexe
cultural edaphoclimatique en Afrique tropicale (Plants indi-
cating the level of fertility of the edaphic climatic cultural complex in Tropical Africa). Agron. Trop. 3. 246-
257. 1948.

1000 PORTERES, R.
L'assolement dans les terres a arachides du Senegal.
Rev. Bot. Appl. 1950, 30:44-50.

1001 PORTERES, R.
Les cereales mineures du genre Digitaria en Afrique et
en Europe. J. Agric. Trop. Bot. Appl. 1955, 2:349-386.

1002 POTLAICHUK, V. I.
Pirikulyarioz prosa (Pyriculariosis of millet).   Zashch.
Rast. Vredit. Bolez. 9(11):51. 1964.

1003 POWELL, J. B.
Reciprocal translocations, Iso-choromosomes, multiploid
microsporocytes, autotriploids, and accessory chromosomes in pearl millet, Pennisetum glaucum.   A talk presented at the 17th Annual meeting of the Georgia section,
American Society of Agronomy in Athens, Georgia, Jan.
15, 1964.

1004 POWELL, J. B.
Cytogenetics of pearl millet (Pennisetum glaucum L.) -
reciprocal translocations, isochromosomes, multiploid
microsporocytes, autotriploids and accessory chromosomes (p. 64). Proc. Ass. S. Agric. Wkrs. , 61st Ann.
Conv. , Atlanta, Ga. , Feb. 3-5, 1964. pp. 299.

1005 POWELL, J. B.
Forage grass cytogenetics and chromosome engineering.
Talk presented at the Genetics Seminar, University of
Fla. , Gainesville, Fla. on 16 March, 1964.

1006 POWELL, J.B.
Cytogenetics in pearl millet improvement. Talk presented at the 21st Southern Pasture and Forage Crops Improvement Conference, University of Fla., Gainesville, Fla., 23 April, 1964.

1007 POWELL, J.B.
Nucleolar organizing accessory chromosomes and their association with nucleolar A-chromosomes in Pennisetum typhoides (p. 76), Annual Meetings, American Society of Agronomy, Kansas City, Mo., Nov. 15-19, 1964, pp. 126.

1008 POWELL, J.B. and BURTON, G.W.
Recovery of miniature accessory chromosomes in mutagen-treated pearl millet among second-generation progeny (p. 50). Proc. Ass. Sth. Agric. Wkrs. 62nd Ann. Conv., Dallas, Texas, 1-3 February, 1965, pp. 286.

1009 POWELL, J.B.
Metaphase I configurations and non-disjunction frequencies of thermal-neutron-induced translocations in pearl millet. Report for U.S. Atomic Energy Commission Contract No. At-(40-1)-2976

1010 POWELL, J.B. and BURTON, G.W.
1964 Annual Report: Grass Breeding Project - Tifton, Georgia.

1011 POWELL, J.B. and BURTON, G.W.
Genetic and cytogenetic effects of recurrent thermal neutron and ethyl methane sulfonate on ten inbred lines of pearl millet (P. typhoides) p. 17. Abstracts of the Annual Meetings of the American Society of Agronomy, Columbus, Ohio, on 31 October - 5 November, 1965, pp. 134.

1012 POWELL, J.B. and BURTON, G.W.
Nucleolus-organizing accessory chromosomes in pearl millet, Pennisetum typhoides. Crop Sci. 6:131-134. 1966.

1013 POWELL, J.B. and BURTON, G.W.
A suggested commercial method of producing an interspecific hybrid forage in Pennisetum. Crop Sci. 6:378-379. 1966.

1014 POWELL, J.B. and BURTON, G.W.
Miniature centric fragment chromosomes in mutagen-treated pearl millet, Pennisetum typhoides. Crop Sci. 6:590-593. 1966.

1015 POWELL, J.B. and TAYLORSON, R.B.
Induced sterility and associated effects of dimethylarsinic acid treatment of pearl millet. Crop Sci. 7(6):670-672. 1967.

1016 POWELL, J.B.
      Standard and non-standard Karyotypes in pearl millet,
      Pennisetum typhoides. Agron. Abstr. Madison 1967. p.
      16.

1017 POWELL, J.B. and BURTON, G.W.
      Polyembrony in pearl millet, Pennisetum typhoides. Crop
      Sci. 8(6):771-773. Illus. 1968.

1018 PRADAT, A.
      The war on animal pests: the problem of Quelen quelea
      quelea (millet eaters). Paper presented at the CCTA/
      FAO Symposium on Savannah Zone Cereals. Dakar, 29
      Aug. to 4 September, 1962.

1019 PRASADA, R.
      Studies on the formation and germination of teliospores of
      rusts. I. Indian Phytopathol. 1:119-126. 1948.

1020 PRIMAK, N.N. and JAKOVLEV, A.G.
      Thermal emasculation of millet. Agrobiologiya 1964, 163-
      164.

1021 PRIMO, A.T.
      Effect of date of maturity on the yield, quality and man-
      agement of pearl millet used for a forage plant. Thesis,
      submitted to Univ. of Ga., in partial fulfillment of the re-
      quirements of the M.S. Degree, Athens, Ga., 1967.

1022 PRINE, G.M. and others
      Physiological responses of Florida forage crops to environ-
      mental variables. Ann. Rep. Fla. Agric. Exp. Sta. 1958.
      50-51.

1023 PROJECT ... (PIRRCOM, New Delhi)
      Annual Report for 1964: Pennisetum. Indian Council of
      Agricultural Research 4:91-95, 158-163 and 226-231.

1024 PRUTHI, H.S.
      Descriptions of some new species of Empoasca Walsh
      (Eupterygidae Jassoidea) from North India. Indian J. Ent.
      2(1):1-10. 1940.

1025 PRUTHI, H.S. and SINGH, M.
      Stored grain pests and their control. III. Natural re-
      sistance of different grains to insect attack. Misc. Bull.
      Imp. Coun. Agric. Res. India 1943, No. 57, p. 13.

1026 PUHALJSKII, A.V.
      All scientific achievements should aid production. Zemle-
      delie 1954. No. 3:89-97.

1027 PULMAN, I. A.
The yield of millet (Panicum miliaceum) in dependence on
meteorological factors. Tr. Sel'. -Kohz. Met. , 1909, No.
5, I. pp. 6-19. 1909. Exp. Sta. Rec. 23: 117.

1028 PUL'NOV, I. V.
On the biology of millet fly Stenodiplosis panici Rohd in
Kuybyshev Region. Uch. Zap. Kuybyshevsk. Gos. Ped.
Inst. 1956(16):121-141. 1956.

1029 PUTTARUDRAIAH, M. and RAJU, R. N.
Observations on the host range, and control of Azazia
rubricans Boisd. Indian J. Ent. 14(2): 158. N. Delhi,
1952.

1030 PUTTARUDRAIAH, M. and CHANNABASAVANNA, G. P.
Host range of Cacoecia Epicyrta Meyrick (Ecn). Indian J.
Ent. 12:267-268. 1950.

1031 QUAGLIARIELLO, G.
The chemical composition of millet flour. Pubbl. Comitato Sci. Aliment R. Accad. Naz. Lincei, No. 11, pp. 19.
1921.

1032 RACHIE, K. O.
The systematic collection of sorghums, millets and maize
in India. Report of the Rockefeller Foundation and the
Indian Council of Agricultural Research, New Delhi. 1965.

1033 RACHIE, K. O.
Utilizing genetic diversity in sorghums and millets im-
provement. International Symposium on "the impact of
Mendelism on agriculture, biology and medicine" held at
IARI, New Delhi on 15 - 20 February, 1965.

1034 RACHIE, K. O.
Progress in pearl millet improvement in India. Unpublished 1966.

1035 RACHIE, K. O.
Some implications of protogyny and cytogenic male sterility in breeding bajra (P. typhoides S. and H). Paper presented at the 53rd Indian Science Congress (Agriculture
Section) held at Chandigarh, Punjab on 7 January, 1966.

1036 RACHIE, K. O.
Sorghum and millet hybrids for India. Span 9:1. 1966.
(Burmah Shell, London).

1037 RACHIE, K.O., SINGH, Amarjit and BAKSHI, J.S.
Development of hybrid grain millets (Pennisetum typhoides
S. & H.) for India. Presented at the Western Branch of
the American Society of Agronomy Meetings on 27-29
June, 1967, at Las Cruces, New Mexico.

1038 RACHIE, K.O., BAKSHI, J.S. and SINGH, Amarjit
Progress in bajra improvement in India. Presented at
the 53rd Indian Science Congress (Agriculture Section) at
Chandigarh, Punjab on 7 Jan. 1966.

1039 RACHIE, K.O. and PALIWAL, R.L.
Controlling birds in experimental crops.   Unpublished,
1966.

1040 RACHIE, K.O. and GUPTA, V.P.
Preliminary investigations with pre-emergence weedicides.
Sorghum Newsletter 7:58-59.   1964.

1041 RAHEJA, P.C.
Water requirements of Indian field crops.   Res. Ser. 28
Indian Coun. Agric. Res. 1961, pp. 25.

1042 RAIMO, H.F. and ROCHA, G. Leme da
Contribuicao para o estudo dos substitutos dos farelos de
trigo na alimentacao das aves. 1. Cereal Adlay. Bol.
Indust. Animal 1950, 11: No. 1-2: 85-95.

1043 RAJ, S.M., MAHUDESWARAN, K. and SHANMUGASUNDA-
RAM, A.
Observation on the hot water technique of emasculation of
ragi flowers (Eleusine coracana Gaertn).   Madras Agr. J.
1964, 51: p. 71.

1044 RAJA, K.V.J. Naidu
Fodder crops in India. ICAR-New Delhi. 1932.

1045 RAJARATHINAM, S. et al.
Pilot scheme for bajra production in Madras.   Indian Fmg.
1965. 15: No. 6: p. 9.

1046 RAJU, D.G. and RAO, P.G.
Field testing of fungicides. 3. Andhra Agric. J. 1961,
8: No. 2:130-134.

1047 RAKHIMBAEV, I.R.
Aminokislotyni sostav belkov prosa kazakhstana (Amino
acid composition of the proteins of Kazakhstan millet).
Tech. Inst. Food Ind., Moscow, USSR. Prkl. Biokhim,
Mikrobiol. 3(1):17-20. 1967.

1048 RAKHIMBAEV, I.
Carotenoids in proso millet grain. 13v. Akad. Nauk.

Kazakh S.S.R. Sci. Biol. 1967. No. 6. 51-53. (Bibl.
12., Ru. Kazakhj Tekhnolog Inst. Pishch. Prom. Moscow.)

1049 RAMACHAR, P. and SALAM, M.A.
Rusts of Hyderabad. J. Indian Bot. Soc. 23:192-196.
1954.

1050 RAMAKRISHNAN, C.V.
Evaluation of nutritive values in bajra hybrids. Personal
communication of 4-1-66 (Unpublished).

1051 RAMAKRISHNAN, K.
Investigations of cereal rusts. II. Uromyces setariae-
italicae (diet) Yosh. Indian Phytopathol. 2:31-34. 1949.

1052 RAMAKRISHNAN, K.V.
Studies on the morphology, physiology and parasitism in
the genus Piricularia in Madras. Proc. Indian Acad. Sci.
B. 27:174-193.

1053 RAMAKRISHNAN, P.S.
Ecology of Echinochloa colonum Link. Proc. Indian Acad.
Sci. Sect. B. 52(3):73-90. 1960.

1054 RAMAKRISHNAN, T.S.
Top rot ('tisted top' or 'pokha boeng') of sugarcane, sorghum and cumbu. Curr. Sci. 10:406-408. 1941.

1055 RAMAKRISHNAN, T.S.
Administrative Report of the Government Mycologist, Madras Agriculture Department, 1951-52. 1952.

1056 RAMAKRISHNAN, T.S.
Diseases of millets. Indian Council of Agricultural Research, New Delhi, pp. 67-142. 1963.

1057 RAMAKRISHNAN, T.S. and SRINIVASAN, K.V.
Additions to the fungi of Madras. XIII. Proc. Indian
Acad. Sci. Sect. B. 36:85. 1952.

1058 RAMAKRISHNAN, T.S. and SUNDARAM, N.V.
Notes on some fungi from South India. III. Indian Phytopathol. 7:64. 1954.

1059 RAMAKRISHNAN, T.S. and SUNDARAM, N.V.
Further studies on Puccinia penniseti Zimm. Proc. Indian
Acad. Sci. Sect. B. 1956, 43:190-196.

1060 RAMAKRISHNAN, T.S.
Additions to the fungi of Madras XII. Proc. Ind. Acad.
Sci. Sect. B. 36(3):111-121. 1952. (Agric. Res. Inst.
Coimbatore, India.)

1061 RAMAN, V. S., NAIR, M. K. and KRISHNASWAMI, D.
The cytology of Pennisetum ruppellii Steud. - a reinvesti-
gation. J. Indian Bot. Soc. 1962, 41:243-46.

1062 RAMAN, V. S., NAIR, M. K., and KRISHNASWAMI, D.
Studies on the interspecific hybrid of Pennisetum typhoides
x P. purpureum. VII. Reciprocal cross-derivatives of
the allotetraploid. J. Indian Bot. Soc. 1963, 42:469-73.

1063 RAMAN, V. S. and NAIR, M. K.
Studies on the interspecific hybrid of Pennisetum typhoides
x P. purpureum. VIII. The cytology of certain derivatives. J. Indian Bot. Soc. 1964, 43:508-514.

1064 RAMANATHAN, M. K. and GOPALAN, C.
Effect of different cereals on blood sugar levels. Indian
J. Med. Res. 45(2):255-262. 1956.

1065 RAMASWAMI, S., SARMA, P. S. and SREENIVASAYA, M.
Studies on insect nutrition-assay of "quality" in crops.
Curr. Sci. 11(2): 53-54. 1942.

1066 RAMASWAMY, C.
Meteorological factors associated with the ergot epidemic
of bajra (Pennisetum typhoides) in India during the Kharif
season 1967. A preliminary study. Curr. Sci. 37(12):
331-335. Illus. 1968.

1067 RAMCHANDRA RAO, S.
Millets in Andra Pradesh. Krishik 1958-59. 6:84-87.

1068 RAMOND, C.
For a better understanding of the growth and development
of Pennisetum millets. Agron. Trop. Paris 1968, 23, No.
8. 844-863. (Biblio. 3 i F. E. E. S.).

1069 RAMULU, K. S.
Meiosis and fertility in derivatives of amphidiploid of
Pennisetum. Caryologia 1968: 21: 147-156.

1070 RAMULU, K. S. and PONNAIYA, B. W.
A study of variation pattern in the progenies of an amphi-
ploid of Pennisetum. Madras Agr. J. 1967, 54. 5030511.

1071 RAMULU, U. S. and KULANDAI, A. Maria
The composition of the ragi (Eleusine coracana) grain and
straw as affected by the application of farmyard manure
and superphosphate fertilizer. Madras Agr. J. 1964, 51:
No. 9: 379-385.

1072 RANE, F. W.
Experiments with roots and forage crops. New Hampshire Sta. Bul. 57. pp. 127-153. Exp. Sta. Rec. 10.

945 and 946. 1898-1899.

1073 RANGARAJAN, A. V.
Arthrodes sp. A new tenebrionid beetle pest of cumbu
(bajra) and its control. Sci. Cult. 31(1):41-43. 1965.

1074 RANGARAJAN, A. V.
A note on the incidence of Holotrichia sp. (Melolonthidae:
Coleoptera) on ragi (Eleusine coracana Gaertn) and its
control. Sci. Cult. 32(12):604-605. 1966.

1075 RANGASWAMI, G. and PRASAD, N. N.
Studies on the survival of plant pathogens added to the
soil. I. Fusarium spp. and Xanthomonas cassiae. Indian Phytopathol. 14(1):83-87. 1961.

1076 RANGASWAMI, G., PRASAD, N. N. and EASWARAN, K. S. S.
Two new bacterial diseases of sorghum. Andhra Agric.
J. 8(6):269-272. 1961.

1077 RANGASWAMI, G., PRASAD, N. N. and ESWARAN, K. S. S.
Bacterial leaf spot diseases of Eleusine coracana (ragi)
and Setaria italica (tennq) in Madras. Indian Phytopathol. 14(1):105-107. 1961.

1078 RANJAN, S. V. G. and RAO, H. G. G.
Effect of micronutrients on crop response and quality in
Mysore State. J. Indian Soc. Soil Sci. 12(4):203-466.
1964.

1079 RAO, C. V. and PURUSHOTHAM, N. P.
Studies on the nutritive value of Echinochloa crusgalli
Beauv. Indian Vet. J. 40(11):714-717. 1963.

1080 RAO, D. G., VARMA, P. M. and CAPOOR, S. P.
Studies of mosaic disease of Eleusine in the Deccan. Indian Phytopathol. 18(2):139-150. 1965.

1081 RAO, D. V. N. and DAMODARAM, G.
Studies on correlation of certain plant characters to yield
in pearl millet (Pennisetum typhoides S. & H.) Andhra
Agric. J. 1964, 11: No. 1: 22-25.

1082 RAO, D. V. N., DAMODARAM, G. and MOORTHY, A. K. R.
Preliminary studies on the proper time of sowing and
plant population in irrigated ragi in Chittoor District.
Andhra Agric. J. 12(6):208-213. 1965.

1083 RAO, I. M.
Bajra crop in southeastern Punjab. Indian Fmg. 5(4):173-
175. 1944.

1084 RAO, I.S., RAO, K.R. and MURTHY, K.K.
     Varietal trial with irrigated ragi in rabi season in sandy
     soils. Andhra Agric. J. 1965, 12, No. 3, 100-5.

1085 RAO, K.P.
     Annual Reports of the Millet Breeding Station, Coimbators
     for the year 1947-1948. pp. 26. 1948.

1086 RAO, K.P.
     The role of millets in increasing food production in Mad-
     ras. Madras Agr. J. 35:311-316. 1948.

1087 RAO, K.P. and NAMBIAR, A.K.
     Manuring of millets in Madras.   Madras Agr. J. 39:73.
     1952.

1088 RAO, K.P.
     Methods to be adopted to maximize production and devel-
     opment of improved strains of millet seeds.   Madras Agr.
     J. 40: 125-128. 1953.

1089 RAO, W.V.B. Sundara-, MANN, H.S., PAUL, N.B. and
     MATHUR, S.P.
     Bacterial Inoculation experiments with special reference to
     Azotobacter. Indian J. Agr. Sci. 33(4):279-290. 1963.

1090 RAO, Y.Y., LATCHANNA, A. and AHMED, M.K.
     Yield and yield components of Hybrid Bajra (HB-1) as in-
     fluenced by nitrogen, phosphorus and potash levels. Indi-
     an J. Sci. Ind., 1967. 1, 19-24. (Coll. Agric. A.P.
     Agric. Univ. Hyderabad - 30).

1091 RATNASWAMY, M.C. and DHANARAJ, L.
     A non-lodging mutant in tenai (Setaria italica Beauv) the
     Italian millet. Sci. Cult. 1961, 27:194-195.

1092 RATNASWAMY, M.C. and PONNAIYA, B.W.X.
     Discriminant function technique in selection for yield in
     Setaria italica Beauv. the Itallian millet.   Madras Agr.
     J. 1963, 50: p. 84.

1093 RAU, N.S.
     On the chromosome numbers of some cultivated plants of
     South India. J. Indian Bot. Soc. 8:126-128. 1929.

1094 RAZVYAZKINA, Mme G.M., KAPKOVA, Mme E.A. and
     BELYANCHIKOVA, Mme Yu V.
     Virus polosatoi mozaiki pshenitsy (Wheat streak millet
     virus). Zashch. Rast. Vredit. Bolez. 8(9):54-55. 1953.

1095 RESTUCCIA, G.
     Investigations over 2 years on the biological behaviour and
     yield of the synthetic variety S.M. Gladiator of Pennisetum

glaucum. Sementi Elette 1966, 12, No. 4, 246-256.

1096 RHIND, J. M. L. C.
Cytogenetic study in Panicum deustum (Research Project
N-Pm. 82). Agric. Res. (Pretoria) 1966: Part I: 333-
334.

1097 RHODESIA AND NYASALAND, Federation of Ministry of Ag-
griculture
Annual Report of the Secretary to the Federal Ministry of
Agriculture, including the Department of Research and
Specialist Services, Conservation and Extension Services
and Veterinary Services, for the year ended 30 Septem-
ber, 1954. Zomba, 1955. pp. 115. (Rhodes Matopos
Estate. Crop Trials, pp. 46-47).

1098 RHODESIA (Northern) DEPT. AGRIC.
Report for the year 1949. Lusaka, N. R., 1950, pp. 19.

1099 RHODESIA (Northern) DEPT. AGRIC.
Annual report for the year 1954. Lusaka, 1955, pp. 28.
(6. Finger millet, p. 28).

1100 RHODESIA (Southern) DEPT. AGRIC.
Annual Report of the Ministry of African Agriculture including the reports of the Departments of Agriculture and
Co-operatives and African Marketing, Northern Rhodesia
for the year 1961 (1962).

1101 RICAUD, R. and STELLY, M.
The effects of phosphorus levels and of phosphorus treatment on the yield and phosphorus content of millet grass
grown on three soils in Louisiana. Proc. Ass. Sth.
Agric. Wkrs. 57:73-74. 1960.

1102 RICHARDS, G. J.
Variations in the proportions and crude protein contents
of leaf blades, leaf sheaths, stems and inflorescences in
Setaria sphacelata. S. Afr. J. Agric. Sci. 1963, 6:221-
232.

1103 RIGGL, E.
Investigations on the influence of different depths of plant-
ing on the growth of cereals. Vrtljschr. Bayer, Landw.
Rat., (1907), No. 2, Suppl., pp. 313-378. 1907. Exp.
Sta. Rec. 19:731. 1907-1908.

1104 RIRIE, D.
Water grass control in sugar beets. Proceedings of the
6th Annual California Weed Conference, 32-33. 1954.

1105 RITCHIE, A. H.
Rep. Dep. Agric. Tanganyika, 1926. pp. 33-36. 1926.

1106 ROBERT, Alice L.
New hosts for three Helminthosporium species from corn.
Pl. Dis. Reptr. 46(5):321-324. 1962.

1107 ROBERTS, W. and KARTARSINGH, S. B. S.
A textbook of Punjab Agriculture. Civil and Military Gazette Ltd. Lahore, 1947.

1108 ROBOCKER, W. C., KERR, H. D. and BRUNS, V. F.
Characteristics and control of swainsonpea. Weeds 1964,
12: no. 3:189-191.

1109 ROCKEFELLER FOUNDATION, THE
Programme in the Agricultural Sciences. Annual Report
1962-63. pp. 256. 1962. New York, 1963.

1110 ROCKEFELLER FOUNDATION, THE
Programme in the Agricultural Sciences. Annual Report
1962-   pp. 205-209. 1963. New York, 1964.

1111 ROCKEFELLER FOUNDATION, THE
Programme in the Agricultural Sciences. Annual Report
1963-1964. pp. 184-189. 1964. New York, 1965.

1112 ROCKEFELLER FOUNDATION, THE
Programme in the Agricultural Sciences, Annual Report
1964-1965. pp. 162-167, New York, 1966.

1113 ROCKEFELLER FOUNDATION, THE
Programme in the Agricultural Sciences, 1965-1966. pp.
152-163. New York, 1967.

1114 ROCKEFELLER FOUNDATION, THE
Programme in the Agricultural Sciences Annual Report
1966-1967. New York, 1968.

1115 ROMANOV, I. K.
Seed and Seed Growing. Selek. Semenovodstvo 1953. No.
4:17-20.

1116 ROONWAL, M. L.
Notes on the bionomics of Hierglyphus nigrorepletus, Bolivar (Orthoptera, Acrididae) at Benares, United Provinces,
India. Bull. Ent. Res. 36(3):330-341. London, 1945.

1117 ROSEN, H. R.
A bacterial disease of foxtail (Chaetochloa lutescens).
Ann. Mo. Bot. Gdn. 9(40):333-402. 1922. Exp. Sta. Rec.
50:349.

1118 ROSTOTSEVA, Z. P.
The responses of millet varieties to changes in temperature conditions. Agrobiologiya 1959, No. 6:883-887.

1119 ROSTOVTSEVA, Z. P.
The nature of the change in the reactions of plants to environmental factors when there is an extension of the range of cultivation of a species. In: Eksperimental'nyi morfogenez. Moskov. Univ. Moscow. 247-256. 1963.

1120 ROSTOVTSEVA, Z. P.
On the morphological and physiological variability of biological forms within a species. Experimentnyi Morfogenez. Moskov. Univ. Moscow. 343-350. 1963.

1121 ROTMISTROV, V.
Distribution of the roots of some annual cultivated plants. Zh. Opyt. Agron. 8(6):667-705, (1907); 9(1):1-24, (1908); 1908-1909.-Exp. Sta. Rec. 1908-1909. 20. 732.

1122 RUDDELL, E. C.
Personal communication on finger millet of 14 February, 1968. Univ. of California (Davis) 1968.

1123 RUSANOV, N. M.
Reclamation of semi-desert steppe-lands. Results of scientific research work at the Karaganda Agricultural Research Station during 1938 1947. Uclek. Demenovdslvo 1948: 15: No. 11:22-31.

1124 SABAH (Malaysia) DEPT. OF AGRICULTURE
Annual Report for the year 1965. Jesselton, 1967 pp. 108.

1125 SABAN IS, T. S.
Studies in Indian millets. Agric. Anim. Husb. India. 1936-1937. 1937.

1126 SABAN IS, T. S.
Studies in Indian millets. Agric. Livestock India 6:506-516. 1936.

1127 SADASIVAN, T. S. and RAO, P. A.
The uptake of nutrients by ragi (Eleusine coracana). Mysore J. Agric. Sci. 1968. 2: No. 2: 133-140. (Bibl. 7: Dir. Chem. and Soils, Coll. Agric. Bangalore, India.)

1128 SADASIVAN, T. S. and SUBRAMANIAN, C. V.
Studies on the growth requirements of Indian fungi. Trans. Brit. Mycol. Soc. 37:426-430. 1954.

1129 SAFEEULLA, K. M. and SHAW, G. G.
Sporangial germination of Sclerospora graminicola and inoculation of Pennisetum glaucum. In: 45th Annual Meeting of the Pacific Division of the American Phytopathological Society, Stanford Univ., California, 1963. Phytopath-

ology 53(10):1239-1240. 1963.

1130 SAFEEULLAH, K.M. and THIRUMALACHAR, M.J.
Periodicity factor in the production of a sexual phase in
Sclerospora graminicola and S. sorghi and the effect of
moisture and temperature on the morphology of the spor-
angiosphores. Phytophath. Z. 26:41-48. 1956.

1131 SAHARNOI
Thirty years' breeding and seed production work on cereals and grain legumes (1922-1952). Vsesojuznyi Naucno-
issledovateljnyi Institut Saharnoi Svekly (Sugar-beet Research Institute) Moskva 1956: pp. 468.

1132 SAHR, C.A.
Report of the Agronomy Department (Univ. of Hawaii).
Hawaii Sta. Rpt. 1915, 39-44. 1951. Exp. Sta. Rec. 35:
528.

1133 SAIKI, T.
Progress of the science of nutrition in Japan: chemical
properties and nutritive value of the protein of Italian
millet (Setaria italica Kth.) Geneva: League of Nations,
Health Organ., 1926, pp. 387 (pp. 331-335). 1926. Exp.
Sta. Rec. 60:290-291.

1134 SAMBASIVA RAO, I., KRISHNAMURTHY, K. and RANGA
RAO, K.
Ragi (Eleusine coracana Gaertn) varietal trials with AKP.
6 and Co 7 in Kharif and rabi seasons in sandy soils.
Andhra Agric. J. 1968. 15: No. 1:7-9.

1135 SAMBASIVA RAO, I. and RAGHAVALU, P.
Studies on transplantation versus direct sowing or ragi
(Eleusine coracana Gaertn) at three different times in
kharif and rabi seasons in sandy soils. Andhra Agric. J.
1964, 11: No. 4:125-136.

1136 SAMPSON, H.O.
Roots, Emp. Cotton Growing Rev. 16:165-170. Madras
Agr. J. 28: 388-392. 1940.

1137 SANDHU, A.S.
Effect of row direction on the growth and yield of bajra
(Pennisetum typhoides). Indian J. Agron. 1964, 9: No. 1:
19-22.

1138 SANKARAM, A.
A plea for more millets. Indian Fmg. 7(2):72-76. 1946.

1139 SANKARAN, S., RAMACHANDRAN, M. and KALIAPPA, R.
A note on the tillering habit of hybrid Bajra-1 (Pennisetum
typhoides S & H) Madras Agr. J. 54(6):320-323. 1967.

1140 SANTELMANN, P.W., MEADE, J.A. and PETERS, R.A.
      Growth and development of yellow foxtail and giant foxtail.
      Weeds 11(2):139-142. 1963.

1141 SARAEV, P.I.
      Experiences in obtaining high yields of millet in the West
      Kazahstan and Aktubensk Territories. Opyt. Agron., No.
      5, 18-23. 1941.

1142 SASTRY, K.S.K., and DAWSON, M.J.
      Presence of an inhibitor in the "seedcoat" of ragi. Curr.
      Sci. 1966, 35, No. 24, 612-613.

1143 SASTRY, K.S.K. and APPIAH, K.M.
      Effect of thiamine on growth of roots of Dolichos biflorus
      and Eleusine coracana. Mysore J. Agric. Sci. 2(2):106-
      110. Illus. 1968.

1144 SAUNDERS, W., et al.
      Field experiments with farm crops. Canada Exp. Farms
      Rpts. 1901, pp. 533. Exp. Sta. Rec. 14: 132. 1902-
      1903.

1145 SAUNDERS, W., et al.
      Field experiments with farm crops. Canada Exp. Farms
      Repts. 1902, pp. 7-373. Exp. Sta. Rec. 15:137. 1903-
      1904.

1146 SAUNDERS, W., et al.
      Field experiments with farm crops. Canada Exp. Farm
      Rpts. 1903, pp. 5-412. Exp. Sta. Rec. 16:249. 1904-
      1905.

1147 SAUNDERS, W., et al.
      Forage plants and cereals at Highmore Sub-station, 1904-
      1905. S. Dak. Sta. Bull. 96. pp. 23-60. Exp. Sta. Rec.
      18: 133. 1906-1907.

1148 SAVILE, A.H., THORPE, J.C., COLLINGS-WELLS, L.J.,
      and PEERS, A.W.
      Notes on Kenya Agriculture. 1. Ceral Crops. East Afr.
      Agr. J. 1958, 23: No. 4:228-233.

1149 SAVOFF, Ch.
      Experiments for controlling loose smut (Ustilago crameri
      Korn of millet. Rev. Inst. Recherches Agron. Bulg. 4
      (3):91-99. 1929.

1150 SAVUR, R.M.
      Correspondence. Madras Agr. J. 43:166. 1956.

1151 SAVZDARG, S.F.
      Agroclimatic indices for yields of proso millet grown under

different intensities of cultivation.   Tr. Tsent. Inst. Prognozov 1965, 140:71-81.

1152 SAWICKI, J.
Investigations on the resistance of millet (P. milicaeum
L.) to smut, S. panici-miliacei (Pers) Bubak. Acta Agrar.
Silv., Krakow: Ser. Roln. 1964, 4:117-49.

1153 SAWICKI, J.
The agricultural value of millet strains resistant to smut.
Acta Agrar. Silv., Krakow: Ser. Agrar. 1968: 8: num-
ber 1, 111-136 (Polish).

1154 SAZANOV, V.I., SCIBRAEV, N.S. and PAHOMOV, A. Ja
Effectiveness of selection methods in producing elite millet seed (Panicum). Izv. Kujbšev. Sel'skohoz. Inst.
(News Kybyšev Agric. Inst.) 1964: 14:192-196; from Referat. Zh. 1965. Abst. 18:55.31.

1155 SCARASCIA, G.T. and D'AMATO, F.
The content of amino acids of the different varieties of
Pennisetum and sorghum of Senegal. pp. 138-151. Qual.
Plant Mater. Veg. 1958, 3/4: 1-610.

1156 SCAUT, A.
Le Coix lachryma-jobi: Composition chimique, digesti-
bilite et valeur energetique pour le porc. (Congo). Bull.
Agric. Congo 1961, 52:265-270.

1157 SCEGLOV, Y.V.
On the sensitivity of millet varieties to the sodium salt of
2,4-D. Trans. Sci-Res. Inst. Fertiliz. Agric. Soil Sci.
1962. No. 39:92-102; from Referat. Zh. 1963, Abst.
9G99.

1158 SCHAAFFHAUSEN, R.
Adlay or Job's tears - a cereal of potentially greater economic importance. Econ. Bot. 1952, 6: No. 3, 216-227.

1159 SCHAEVERBEKE, J.
Action de la gibberelline sur l'allongement des filets
staminaux chez les graminees. C.R. Hebd. Seanc. Acad.
Sci., Paris 1960, 251: No. 11:1176-1178.

1160 SCHAEVERBEKE, J.
Study of some cytologic and metabolic changes in the sta-
men filaments of grasses during anthesis. C.R. Hebd.
Seanc. Acad. Sci., Paris 1964, 259: No. 22: 4118-4121.

1161 SCHILLING, R.
Groundnuts intercropped with cereals. Oleagineux 20(11):
673-676. 1965.

1162 SCHREIBER, N. W.
Development of giant foxtail under several temperatures
and photo-periods. Weeds 13(1):40-43. 1965.

1163 SCHULTZE-MOTEL, J. and KRUSE, J.
Spelt (T. spelta L.) other cultivated plants and weeds
found in Central Germany in the Iron Age (early). Kulturpflanze 1965, 13:586-609.

1164 SCHUMAKER, G.
E. A. Agriculture and Forestry Organization. Record of
research from 1 January to 31 December, 1966. Ann.
Report 1966 (Nairobi) 1967. p. 188.

1165 SCIBRAEV, N. S. and PAHOMOV, A. Ja
Brief results of millet breeding. Izv. Kujbyšev. Sel'skohoz. Inst. (News Kybysev Agric. Inst.) 1964:184-191:
from Referat. Zh. 1965. Abst. 18.55.30 (Russian)

1166 SEED WORLD
Two new forage grasses developed for Southeast (USA).
Seed World 83(7):15. 1958.

1167 SELVIG, C. G.
Report of the field crops work at the Crookston Substation,
1917. Minn. Sta. Rpt. 1918. pp. 75-78, 79-81. 1917.
Exp. Sta. Rec. 40:733.

1168 SEMENIUK, G. and MANKIN, C. J.
Occurrence and development of Sclerosphora macrospora
on cereals and grasses in South Dakota. Phytopathology
54(4):409-416. 1964.

1169 SEN, A. C.
Basic factors for forecasting epidemic outbreaks of the
rice bug (Leptocorisa varicornis F.). Indian J. Ent. 17
(1):127-128. N. Delhi, 1955.

1170 SENE, D.
Production of millets, sorghums and maize in the Republic of Senegal. Sols. Afr. 1966:11(1-2) 285-290.

1171 SENEGAL-IRAT
Annales du centre de recherches agronomiques de Bambey
au Senegal. Annee 1949. Annales of the centre for agronomic research, Bambey, Senegal, 1949. Bull. Agron.
Minist. Fr. d'Out. Mer. No. 5 1949. Nogent-sur-Marne,
1951.

1172 SEVCENKO, F. P.
Increased resistance to diseases of the seed of spring
crops from sowing in late autumn. Agrobiologiya No. 6:
152-155. 1949.

1173 SEVRYUKOVA, L. F.
The effect of vernalization on an increase in the resistance of millet to smut. Tr. Khar'kovsk. Sel'skokhoz. Inst. 38 (75):142-147. 1962.

1174 SEVRYUKOVA, L. F.
The role of superphosphates in treating millet against smut and the mechanism of the curative action (From: Rodigin, M. N. "Voprosy immuniteta i ozdorovleniya rastenii"). Tr. Khar'kovsk. Sel'Skokhoz. Inst., 43, Kiev, Izdatel'stvo Urozhai, 1964.

1175 SHAH, H. C. and MEHTA, B. V.
Effects of fertilizers on the mineral contents of pearl millet, Pennisetum typhoideum straw. Indian J. Agric. Sci. 1960, 30: 115-128.

1176 SHANKAR, K., AHLUWALIA, M. and JAIN, S. K.
The use of selection indices in the improvement of a pearl millet population. Indian J. Genet. Pl. Breed. 1963, 23: 30-33.

1177 SHANMUGHASUNDARAM, A., SAMATHUVAM, K., PREM-SEKAR, S., and MICHAL RAJ, G.
A new ragi (Eleusine coracana) strain Co. 8. Madras Agr. J. 1965, 52: p. 83.

1178 SHANMUGASUNDARAM, A., RAJASEKARAN, R., SELVARAJ, S. and RAMAMURTHY, T. G.
Evaluation of sugary disease and green ear in cumbu varieties. Madras Agr. J. 1968. 55:37-38.

1179 SHANMUGASUNDARAM, A., RAJASEKARAN, R., SELVARAJ, S. and RAMAMURTHY, T. G.
Note on a new dwarf variety of cumbu. Madras Agr. J. 1968. 5:145-146.

1180 SHARMA, B. B.
Smut disease of 'Sawan,' Echinochloa frumentacea Link caused by Ustilago paradoxa Syd. and Butl. Proc. Nat. Acad. Sci. India, 1963, 33:618-630.

1181 SHARMA, Y. P.
Preliminary report on breeding hybrid Pennisetum. Balwant Vidyapeeth J. Agric. Sci. Res. 1961, 3: No. 2:62-68.

1182 SHCHEGLOV, Yu V.
A chemical method for weed control in millet fields (In: The use of herbicides and plant growth stimulators). Byul. Akad. Nauk Belorussk. SSR: Minsk. 64-69. Referat Zh. Biol., 1962, No. 196478.

1183 SHEN, S. T. and ONG, T. T.
      Effect of selfing upon yield of millet.   Proceedings of the
      1st Plant Breeding Conference, China, 1934, 24-25.

1184 SHEPPERD, J. H. and TEN EYCK, A. M.
      Grain and forage crops.   N. Dak. Sta. Rpt. 1900, pp. 59-
      97. Exp. Sta. Rec. 13:336.   1901-1902.

1185 SHEVCHENKO, I. S.
      Vospriimchivost' Kukuruzy k vozbuditelyu puzyrchatoi gol-
      ovni kurinogo prosa (Susceptibility of maize to the causal
      agent of head smut of barnyard millet).   Tr. Volgograd
      Sel'. -Khoz. Inst. 16:226-238.   1964.

1186 SHLEIFER, S. V. and KUPERMAN, F. M.
      Effect of the spectral composition of light on the develop-
      ment of millet as a function of the length of the photo-
      periods and the alteration of strong and weak illumination.
      Vest. Moskov. Univ. Ser. VI. Biol. Pochvoved. 6. 33-
      42. 1965.

1187 SHULOV, I. S.
      Various smaller experiments with fertilizers and soils.
      Izv. Moskov. Selsk. Khoz. Inst. 15(1):116-125.   1909.
      Exp. Sta. Rec. 22:223.

1188 SHUMKOVA, M. N.
      Itogi Raboty po selektsii prosa (Russian: Breeding proso).
      Tatarskaya Respublika Gosudarstvennaya Sel'skokhozyaist-
      vennaya Opytnaya Stantsiya, Trudy 1: 105-120.   1961.

1189 SIKKA, S. M. and JOSHI, A. B.
      Intensification of research on the genetic improvement of
      millets.   Proc. of the Conference of Workers on Millets
      held at Kolhapur 1955.

1190 SILL, W. H. and AGUSIOBO, P. C.
      Host range of the wheat streak-mosaic virus.   Pl. Dis.
      Reptr. 39:633-642.   1955.

1191 SILVA, Pereira Abelda
      Millet bread.   Some scientific aspects of its home baking.
      179 p. Univ. Port. 1942.

1192 SIMMONDS, J. H.
      Report of the plant pathology section.   Qd. Dep. Agr.
      Stock Ann. Rpt. 1947-1948. pp. 33-35. 1948.

1193 SIMONCELLI, Fausta Lintas
      Millet for bread-making.   Rend. Ist. Superiore Sanita
      (Rome) 10(5):813-823.   1947.

1194 SIMPSON, C. E.
      An investigation of the artificial manipulation of methods
      of production through interspecific hybridization in Penni-
      setum. Diss. Abstr. 1968: 28: Order No. 68-5023:
      3957B-58B (Abst.).

1195 SINGH, D.
      Smaller millets are in the news.   Indian Fmg. 1957.   6:
      No. 11: p. 15.

1196 SINGH, D. N.
      Supernumerary chromosomes in some grasses.   Caryologia
      18(3):547-553. 1965.

1197 SINGH, D. N.
      A note on pollen grain mitosis in Eleusine indica (Linn.)
      Gaertn.   Sci. Cult 31(6): 306-307.   1965.

1198 SINGH, D. N.
      A naturally occurring dicentric chromosome in Eleusine
      coracana (Linn) Gaertn.   Curr. Sci. 1965, 34:153-154.

1199 SINGH, D. N. and GODWARD, M. B. E.
      Cytological studies in the gramineae.   Heredity 1960, 15:
      193-197.

1200 SINGH, Gian and ATHWAL, D. S.
      Variability in Kangni-2. Genotype x environment interac-
      tion, heritability and genetic advance.   Indian J. of Genet.
      Pl. Breed. Vol. 26(2):153-161.   July 1966.

1201 SINGH, H. and PUSHPAVATHY, K. K.
      Morphological and histological changes induced by Sclero-
      spora graminicola (Sacc.) Schroet. in Pennisetum ty-
      phoides Stapf et Hubb. Phytomorphology 15(4):338-353.
      1965.

1202 SINGH, Manohar, M. and MATHUR, M. K.
      Effect of succinic acid treatment on the performance of
      pearl millet (Pennisetum typhoides S. & H.) Advancing
      Frontiers Pl. Sci. 1966, 17, 143-147. (Bibl. 3 Central
      Arid Res. Inst. Jodhpur, India, seen Jan. 1969.)

1203 SINGH, R. S. and RAJENDRA GROVER, K.
      Sooty mould of cotton and other hosts caused by Micro-
      xyphiella hibiscifolia in North India.   Pl. Dis. Reptr.
      52(8):602-604. Illus. 1968. (P. typ).

1204 SINGH, S. G. and SINGH, S. R.
      Bajra Hybrid 1 effects a revolution.   Indian Fmg. 16(2):
      12-13. 1966.

1205 SINGH, S. P. and SHARMA, Y. P.
White Bajra. Sci. Cult. 1964, 30: No. 5:228-239.

1206 SINGH, V. and SINGH, M.
Effect of different levels of nitrogen with and without phosphorus on yield and quality of bajra (Pennisetum typhoides) fodder. Indian J. Sci. Ind. 1968, 2, No. 1 - 23-6. (Bibl. 4. Dept. Agron. Govt. Agric. Coll. Kanpur, India).

1207 SINHA, S. and DALELA, G. G.
Effect of certain chemicals on the rust development on bajra (Pennisetum typhoides). Symposium on the role of therapeutic treatments for controlling plant diseases. Indian Phytopathol. 16(1):104-111. 1963.

1208 SINHA, S., and KAPOORIA, R. G.
An aspect of microbiol control of bajra (Pennisetum typhoides) rust (Puccinia penniseti) Bull. Indian Phytopath. Soc. 3:61-64. 1966.

1209 SIRIUSOV, M. G.
Classification of broom millets. Selsk. Khoz. i Lesov., 246 (1914) Dec., pp. 556-573. 1914. Exp. Sta, Rec, nn nn4.

1210 SKARDUN, D. S.
The choice of Setaria varieties and certain questions of agronomy on peat soils. Trans. Sci.-Res. Inst. Agric. Cent. Dist. Non-Chernozem Zone 1963, No. 19, 82-88.

1211 SKODENKO, V. I.
The question of increasing the resistance of millet to smut. pp. 351-355.

1212 SLONOV, L. S.
The effect of trace elements on some physiological reactions under saline conditions (In: Notes on the Fourth Scientific Conference of Graduate Students of Rostov University). Rostov-on-Don 243-245. 1962. Referat. Zh. Biol., 1963. No. 14G36.

1213 SLYKHUIS, J. T.
An international survey for virus diseases of grasses. FAO Plant Protect. Bull. 10(1):1-16. 1962.

1214 SMALL, W.
Diseases of cereals in Uganda. Cir. Dep. Agric. Uganda 8. pp. 19. 1922.

1215 SMALL, W.
Diseases of cereals in Uganda. Cir. Dep. Agric. Uganda 8. pp. 48-57. 1922.

1216 SMETANKOVA, M.
        Relationship between the mineral uptake and dry matter
        yield of overground parts of Panicum miliaceum in pot
        experiments. Ust. Ved. Inf. MZLVH Rostl. Vyroba
        1965, 38:909-926.

1217 SMIRNOV, B. M. and MATVEENKO, G. A.
        Chemical treatment of millet and sorghum.   Zashch.
        Rast. Vredit. Bolez. 8(12):29. 1963.

1218 SMITH, D. T. and CLARK, N. A.
        Effect of soil nutrients and pH on nitrate nitrogen and
        growth of pearl millet (Pennisetum typhoides (Burm.) Stapf
        & Hubb.) and sudangrass (Sorhus sudanense (Piper) Staph.
        Agron. J. 60(1):38-40. 1968.

1219 SMITH, K. M.
        Textbook of plant virus diseases. pp. vi + 615. J. and
        A. Churchill, London, 1957.

1220 SMITH, Roy J., Jr.
        Control of grass and other weeds in rice fields.   Arkan-
        sas Agr. Exp. Stn. Rep. Ser. 167:1-40, 1968.

1221 SNYDER, H. and HUMMEL, J. A.
        The digestibility of hog millet.   Minn. Sta. Bul. 80, pp.
        178-180. Exp. Sta. Rec. 14:993. 1902-03.

1222 SOKOLOV, G. A.
        Creation of green belts in the sandy deserts of Kazahs-
        tan. Nature, 1952, No. 7:69-72.

1223 SOLPICO, F. O. and YAMBAO, A. N.
        Performance test of millet at the Economic Garden, Los
        Baños, Laguna. Philipp. J. Pl. Indust. 1966. 31:219-
        229.

1224 SONAVANE, K. M.
        Longevity of crop seeds. Agric. J. India 23:271-276.
        1928.

1225 SOULE, A. M. and VANATTER, P. O.
        Experiments with oats, millet and various legumes.   Vir-
        ginia Sta. Bul. 168, pp. 2610290. 1907. Exp. Sta. Rec.
        19:532. 1907-1908.

1226 SOUNDARAPANDIAN, G., MENON, P. Madhava and PON-
        NAIYA, B. W. X.
        Heterosis in pearl millet-effect of nitrogen fertilization
        on hybrids. Madras Agr. J. 1964, 51: p. 356.

1227 SOUTH AFRICA, POTCHEFSTROOM
        Weidingnovorsing in Suid-Africa. Progress Reports, 1951-
        1952.

1228 SOUTH AFRICA, POTCHEFSTROOM
      Annual Report of the activities of the Highveld Region,
      Potchefstroom, for the year 1959/1960. South Africa
      Ministry of Agric. Highveld Annual Report 1959/1960.
      pp. 42.

1229 SOUTH ARABIA, DEPT. OF AGRICULTURE & IRRIGATION
      Report for 1964/1965 and 1965/1966. Aden 1966. p. 69.

1230 SPRAGUE, H.B., FARRIS, N.F., CURTIS, N.J. and COLBY,
      W.G.
      Annual hay crops. New Jersey Sta. Bul. 540, pp. 23.
      1932.

1231 SPRAGUE, R.
      Diseases of cereals and grasses in North America. pp.
      xvi + 538. Ronald Press Co., New York. 1960.

1232 SPRING, F.G. and MILSUM, J.N.
      Notes on the cultivation of ragi (Eleusine coracana) Agr.
      Bul. Fed. Malay States 7(3): 154-161. 1919. Exp. Sta.
      Rec. 42:35.

1233 SRINIVAÚA, Iyengar Ó. and GOPALA, Iyengar
      K. Kempu, an improved variety of the Italian millet-Na-
      vane. Mysore Agric. J. 1957, 32, No. 1, 54-56.

1234 SREERAMULU, K.
      Studies on the cytomorphology of the progeny of a raw
      allopolyploid in Pennisetum. Madras Agr. J. 1965, 52:
      p. 362.

1235 SRIDHARAN, C.S., SITARAMAN, P. and THANDAVARAYAN,
      K.
      A note on the performance of cumbu (bajra) selections in
      Tindivanum Taluk of South Arcot District. Madras Agr.
      J. 1968: 55: 91-92.

1236 SRINIVASAN, S.V.
      Some new hosts for Striga. Curr. Sci. 16:320-321.
      1947.

1237 SRINIVASAN, V.K., DATTA, N. and KHANNA, P.O.
      Foxtail millet has a bright future. Indian Fmg. 1965.
      15: No. 5. p. 16.

1238 SRINIVASAN, V.K., DATTA, Mrs. N. and KHANNA, P.O.
      Studies in minor millets, Div. of Botany, Indian Agric.
      Res. Inst., Delhi-12, Unpublished 1965.

1239 SRINIVASAN, V.K. and KHANNA, D.P.
      Plant type and harvest index in ragi (Eleusine coracana
      Gaertn). Curr. Sci. 1967, 36, 49-50.

1240 STANDEL', M.
Setaria in Western Siberia. Nauk. Pered. Opyt. Sel'.
Khoz. 1958. No. 12: p. 65.

1241 STANISZKIS, W.
The phosphorus metabolism in plants. Bull. Intl. Acad.
Sci. Cracovie, Cl. Math. et Nat., 6:95-123. 1909. Exp.
Sta. Rec. 22:531.

1242 STAPF, O.
Flora of Tropical Africa. Ed. Sir David, Prain Thisle-
ton-Dyer Gramineae 9:954-1090. 1934.

1243 STEENBOCK, H., SELL, M.T. and JONES, J.H.
Fat soluble vitamin, XI-XIII. XII. The fat soluble vitamin content of millets (pp. 345-354). J. Biol. Chem.
56(2):327-373. 1923. Exp. Sta. Rec. 50: 364.

1244 STEPHENS, D.
Fertilizer trials on cotton and other annual crops in small
farms in Uganda. Exp. Agric. 1968, 4, No. 1. 49-59
(Bibl. 19. Kawanda Res. Sta. Uganda).

1245 STEPIN, V.S.
Etiopatogenez pri otravlenii ovec prosom (Etiology and
pathogenesis of poisoning of sheep by millet). Veterinariya, Moscow 1965, No. 9:61-63.

1246 STEVENS, F.L. and HALL, I.G.
Three interesting species of Claviceps. Bot. Gaz. 50:
460-463. 1910.

1247 STEVENSON, B.
True strain of Big German millet sought by Asgrow
Breeding programme. Sth. Seedsman 1953, 16: No. 6:
p. 63.

1248 STEVENSON, J.A. and JOHNSON, A.G.
The nomenclature of broomcorn millet fungus. Phytopathology 34:613. 1944.

1249 STRONA, I.G.
Agrobiologichne obgruntuvannya strokiv sivbi chumizi.
(The Agro-biological basis for sowing times of foxtail
millet.) Tr. Inst. Genet. i. Selektsii 1955(4):127-136.
Referat. Zh., Biol., 1956. No. 71963.

1250 STUART, W.S.
Formalin as a preventive of millet smut. Indiana Sta.
Rpt. 1900, Exp. Sta. Rec. 13:56. 1901-1902.

1251 SUBBIAH, K.C.
Effect of x-rays on chiasma frequency in Coix aquatica

Royb. (pp. 89-90). Proc. XI Int. Conf. Genet., Hague, Neth., Sept. 1963. 1: pp. 332.

1252 SUBRAHMANYAM, T. V.
The jola grasshopper or Deccan grasshopper (Colemania sphenaroides, Boliv.). Mysore Agric. Call. Yb. 1941-1942. pp. 27-28. Bangalore, 1941.

1253 SUBRAMANIAN, C. V.
Foot rot disease in wheat. Curr. Sci. 31(2):46-48. 1962.

1254 SUBRAMANIAM, C. V. and RAMAKRISHNAN, K.
The fungi of India - a second supplement. J. Madras Univ. B. 22:1-56. 1952.

1255 SUBRAMANIAN, T. R., SANTHANARAMAN, T. and VIJAYA-RAGHAVAN, S.
A serious attack of Sitotroga cerealella Oliv. on standing crops of cholam and ragi at Coimbatore. Curr. Sci. 28 (3):127. 1959.

1256 SUDAN, Centre at M'Pesoba
Ri çut of the Colunlation Centre at M'Pesoba (1952).
Agron. Trop. Nogent, March-April, 1952.

1257 SUFIAN, S. and PITTWELL, L. B.
Iron content of teff (Eragrostis abyssinica) J. Sci. Food Agric. 1968. 19. 439.

1258 SUIKOVSKII, Z.
The effect of trace elements on the pigment system and photosynthesis and their localization in plants. In: Mikroelementy v sel'skom Khozyaistve; Meditsine. Gossal Khozizdat Ukr. SSR: Kiev. 202-205. 1963.

1259 SULAIMAN, M., LUKADE, G. M. and DAWKHAR, G. S.
Effect of some fungicides and anti-biotics on sclerotial development and germination of ergot on Pennisetum typhoideum. Hindustan Antibiot. Bull. 9(2):94-96. 1966.

1260 SULLIVAN, C. Y., BIGGS, W. A., EASTIN, J. D., CLEGG, M. D. and MARANVILLE, J.
Heat and drought studies of sorghum, millet and corn. The Physiology of Yield and Management of Sorghum in Relation to Genetic Improvement. Annual Report No. 2, pp. 14-31. Univ. of Nebr. and CRD-ARS, USDA. April, 1968.

1261 SULLIVAN, E. F.
Performance of supplemental summer forage crops in Pennsylvania, 1957-1959. Prog. Rep. 224 Pa. Agric. Exp. Sta. 1960. pp. 8.

1262   SULLIVAN, E. F. and BUSH, H. L.
       Probability analysis of herbicide response. J. Amer.
       Soc. Sugar Beet Technol. 13(8):721-726. 1966.

1263   SULYNDIN, A. F.
       Hybrids between Setaria italica subsp. maxima and S.
       italica subsp. moharium. Selek. Semenovodstvo 1952, No.
       5:34-36.

1264   SUMKOVA, M. N.
       The results of research on millet breeding. Proc. Tar-
       tar Agric. Res. Sta. 1961. No. 1. 105-20; from Referat.
       Zh. 1961, No. 20. Abst. 20G426: p. 49.

1265   SUMKOVA, M. N.
       Some problems of millet breeding. Selek. Semenovdstvo
       1961, No. 3:36-39.

1266   SUMMANWAR, A. S. and BHIDE, V. P.
       Bacterial red-stripe disease of sugarcane Saccharum offi-
       cinarum caused by Xanthomonas rubrilineans var. indicus
       in Maharashtra. Indian J. Sugarcane Res. Dev. 6(2):65-
       68. 1962.

1267   SUNDARARAMAN, S.
       Ustilagao crameri Koern on Setaria italica Beauv. Bull.
       Agric. Res. Inst. Pusa 97. p. 11. 1921.

1268   SURKOV, V. A.
       On the natural stages of the morphogenesis of grasses.
       Ukr. Bot. Zh. 22(2):47-55. 1965.

1269   SURYANARAYANA, D.
       Quoted in Vasudeva, R. S. and Suryanarayana, D. 1955)
       1954.

1270   SURYANARAYANA, D.
       Green ear disease of bajra and the downy mildew of jowar
       in India. Uttara Bharati (Univ. of U. P.). 6:139-142.
       1959.

1271   SURYANARAYANAN, S.
       Growth factor requirements of Piricularia spp. and Scle-
       rotium oryzae. Proc. Indian Acad. Sci. Sect. B. 58:
       154-188. 1958.

1272   SWAMINATHAN, M. S. and NATH, J.
       Two new basic chromosome numbers in the genus Penni-
       setum. Nature 178:1241. 1956.

1273   SWAZILAND. DEPT. OF AGRICULTURE
       Annual Report of the Research Division 1965/1966 (Mal-
       kerns). 1967. p. 127.

1274 SWIETOCHOWSKI, B. and SUN-JUN-LI
The effect of soil reaction on the growth of cockspur panic-
grass (Echinochloa crusgalli 1/BP). Pam. Pulawski No.
1:155-167. 1961.

1275 SYDOW, P. and SYDOW, H.
Monographia Uredinearum 2:339-340. 1910.

1276 SYME, P. S.
Millet. New Zealand J. Agr. 72(3):117, 119-121. Feb.
1946.

1277 SZEPESSY, I.
Fenyviszonyok hatasa a noveny allenallokepessegenek kialaku-
lasaban. (Effects of light conditions on the development of
disease resistance in plants.) Novenytermeles 10(4):345-
350. 1961.

1278 TABOR, P.
Brown top millet. Agron J. 1951. 43. 100. (U.S. Soil Cons.
Service, Spartanburg, S.C.).

1279 TAIRA, H.
Studies on amino acid contents in food crops. 9. Amino
ша ій і тиіа lшыт hï ргиıшіı іrїuшtıuшв ві ва̂t (Avena sativa) and
ragi (Eleusine coracana). J. Jap. Soc. Food Nutr. 1965.
18:194-196.

1280 TAIRA, H.
Amino acid composition of different varieties of foxtail millet (Setaria italica). J. Agric. Food Chem. 1968, 16, No.
6, 1025-1027. (Bibl. 22) Food Res. Inst. Hamazono-cho,
Fukagawa, Koto-ku, Tokyo, Japan.

1281 TAKAHASHI, N.
Genetical studies with Setaria italica. Jap. J. Genet.
1942, 18:150-151.

1282 TAKAHASHI, N. and HOSHINO, T.
Natural crossing in Setaria italica (Beauv). Proc. Crop.
Sci. Soc. Japan 13:337-340. Jap. J. Bot. 12. 1934.

1283 TAKAHASHI, N. and TAKAHASHI, Y.
Studies on successive cultivation of Italian rye-grass and
wild Echinochloa crus-galli in a rice field converted to an
upland field. On mowing interval, Italian ryegrass variety, and amount of fertilizer. J. Jap. Soc. Grassland
Sci. 1966, 12, 67-73.

1284 TAKASUGI, H.
On the life history, pathogenicity and physiologic forms
of Sclerospora graminicola (Sacc.) Schroet. (Studies in
Nipponese peronsporales. III). J. Agric. Exp. Sta.
Tokyo 2:345-366. (Cf. Rev. Appl. Mycol. 13:629).
1934.

1285 TAKASUGI, H.
The relation of environmental factors and the treatment
of oospores to the infection by oospores of Sclerospora
graminicola (studies in Nipponese peronosporales, IV).
J. Agric. Exp. Sta. Tokyo 2:459-480. 1935.

1286 TAKASUGI, H. and AKAISHI, Y.
Studies on the downy mildew of Italian millet in Manchuria (about the germination of oospores). Res. Bull. S.
Manchuria Rly. Co. 11:1-20. (Cf. Exp. Sta. Rec. 70:
489-490, 1934). 1933.

1287 TALANOVA, M. F.
Feeding value of millet. Zhivotnovodstvo 1957. No. 2:
64-65.

1288 TALATI, N. R. and MEHTA, B. V.
Effect of deep and shallow ploughing on nutrient status of
soil and uptake of nutrients by pearl millet on goradu
soil of Anand. J. Indian Soc. Soil Sci. 1965. 13:143-148.

1289 TALIAFERRO, C. M.
Genetics of reproduction in Pennisetum ciliare. Diss.
Abstr. 1966, 27, Order No. 66-6534, p. 49.

1290 TAMS, W. H. T. and BOWDEN, J.
A revision of the African species of Sesamia guenee and
related genera (Agrostidae-Lepidoptera). Bull. Ent. Res.
43(pt. 4):645-678. London, 1953.

1291 TANAKA, I. and ITO, S.
Phytophthora macrospora (Sacc.) Ito and Tanaka, on
wheat plant. Ann. Phytopath. Soc. Japan 10:126-128.
1940.

1292 TANAKA, Y.
On the starch of glutinous rice and its hydrolysis by dias-
tase. J. Ind. Engng. Chem. 4(8):578-581. 1912. Exp.
Sta. Rec. 28:407.

1293 TANGANYIKA AGRIC. CORP.
Report and Accounts for the period 1 April to 30 September, 1955. Tanganyika Agric. Corp. 1955 (1956): pp. 83.

1294 TASKER, P. K. , RAO, M. N. and SWAMINATHAN
Supplementary value of a processed protein food based on
a blend of coconut meal, groundnut flour and bengal gram
flour to poor Indian diets based on different cereals and
millets. J. Nutr. Diet. 1964, 1:95-97.

1295 TAYLORSON, R. B.
Control of seed production in three annual grasses by dimethylarsinic acid. Weeds 14(3):207-210. 1966.

1296 TAYLOR, T. H. C.
Lygus simonyi, Reut., as a cotton pest in Uganda. Bull.
Ent. Res. 36(2):121-148. London, 1945.

1297 THERON, E. P.
Eragrostis curvula. Fmg. S. Afr. 1968, 43, No. 11, 41-
42. (Cedara, S. Africa).

1298 THERON, J. J.
Annual report of the Secretary for Agriculture in South
Africa for the Year ended 31 August, 1955. (p. 170).
Fmg. S. Afr. 31(359):40-175. 1956.

1299 THOMAS, A. S.
Food crops as indicator plants in Uganda. East Afr. Agr.
J. 1943. 8:136-140.

1300 THOMAS, C. C.
Coix smut. Phytopathology 10(6):331-333. 1920.

1301 THOMAS, K. M.
Detailed Administrative Report of the Government Mycologist, Madras, 1939-1940. 1940,

1302 THOMAS, K. M.
Detailed Administrative Report of the Government Mycologist, Madras, 1940-1941. 1941.

1303 THOMAS, K. M., RAMAKRISHNAN, T. S. and SRINIVASAN,
K. V.
The occurrence of ergot in South India. Proc. Indian
Acad. Sci. Sect. B. 21:93-100. 1945.

1304 THOMPSON, M. J.
Report of field crops work at the Duluth Substation, Minne-
sota, 1918. Minn. Sta. Rpt. 1919. pp. 87-89. 1919.

1305 TIEMANN, A. and KAEMPFIER, E.
Further experience in the growing of millets. Mitt.
Landw. 59. 504-506. 1944.

1306 TIEMANN, A. and KAEMPFFER, E.
The millets. 67 p. Reichsnahrstandsverlag: Berlin.
1947.

1307 TIWARI, M. M. and ARYA, H. C.
Studies on the epidemiology and host parasite relationship
of Selerospora graminicola (Sacc) Schroet. on bajra (Pennisetum typhoides) Stapf. Dept. Bot. Univ. Rajasthan,
Jaipur, India. Bull. Indian Phytopathol. Soc. 3:42-45.
1966.

1308 TIWARI, M.M. and ARYA, H.C.
      Growth of normal and diseased Pennisetum typhoides tis-
      sues infected with Sclerospora graminicola in tissue cul-
      ture. Indian phytopathol. 20(4):356-368. 1967. (Rec'd
      1968).

1309 TIWARI, M.M. and ARYA, H.C.
      Sclerospora gramminicola axenic culture. Science 163
      (3864) 291-293. Illus. 1969.

1310 TONGIORGI, E.
      Grano, miglio e fave in un focolare rituale dell' eta del
      bronzo a grotta misa (bassa valle della fiora). Nuovo
      Gior. Bot. Ital. 1947. 54:804-806.

1311 TOPORKOV, S.
      Combating smut of cereals Zh. Opyt. Agron. 4(1903),
      No. 1, pp. 58-64. Exp. Sta. Rec. 15:49. 1903-1904.

1312 TOTHILL, J.D.
      Agriculture in the Sudan. Oxford, England: Oxford U.
      Press, pp. 1-971. 1948.

1313 TOURTE, R. and BONLIEU, A.
      Culture mecanique-battage et mouture des mils et sor-
      ghos. Colloque CCTA/FAO Sur Les Cereales des Zones
      de Savane. Dakar, 29 Aug. to 4 Sept. 1962. Bureau des
      Publications, Watergate House, London, W.C.2.

1314 TRET'YAKOV, R.V.
      The secretion of water and restoration of turgor by the
      leaves of cut plants following profound wilting as an indi-
      cation of the drought resistance of plants. Sb. Tr. As-
      piratov Molodykh Nauch. Sotrudnikov Vses. Inst. Raste-
      vodstva. 5(19):217-224. 1964.

1315 TROTTER, I.P.
      Millet for forage and grain. Mo. Ag. Exten. Serv. Leaf-
      let 41. 1937.

1316 TSYBUL'KO, V.S.
      Variation of the amount of assimilation products and photo-
      periodism of plants. Fiziol. Rast. 1965, 12, No. 4, 622-
      30.

1317 TUMANOV, I.I. and KONDO, I.N.
      Withering and drought resistance. Proceedings of the
      All-Russian Congress of Botanists, Leningrad. 1928: 57-
      58. 1928.

1318 UGANDA, DEPT. OF AGRICULTURE
Annual Report of the Department of Agriculture, Uganda
Protectorate for the period 1 July, 1945 - 31 March, 1946.
1947: pp. 92.

1319 UGANDA, DEPT. OF AGRICULTURE
Record of Investigations No. 3 for the period 1 April,
1950, to 31 March, 1952, Entebbe, 1953.

1320 UGANDA, DEPT. OF AGRICULTURE
Annual Report for the year ending 31 December, 1956.
Uganda Department of Agriculture Annual Report, 1956,
pp. 62.

1321 UGANDA, DEPT. OF AGRICULTURE
Annual Report for the year ending 31 December, 1958,
Uganda Department of Agriculture Annual Report. 1958,
pp. 75.

1322 UGAROV, A. and others
Melilotas alba in the Irkutsk Province, Sel'hoz. Sibiri
(Agric. Liberia) 5(4).91-95. 1960.

1323 ULLSTRUP, A. J.
Crazy top of some wild grasses and the occurrence of the
sporangial stage of the pathogen. Pl. Dis. Reptr. 39:
839-841. 1955.

1324 UNITED NATIONS
UNRRA European Regional Office, Division of Operation-
al Analysis, Economic Rehabilitation in the Ukraine, Op-
erational Analysis Paper No. 39. pp. 79. UNRRA, London, 1947.

1325 UNITED STATES DEPARTMENT OF AGRICULTURE
Experiment Station Records, vol. 1-95. 1889-1946.

1326 UNITED STATES DEPARTMENT OF AGRICULTURE
Yearbook of Agriculture - 1898. USDA, Washington, D.C.
1898. pp. 272, 276, 278, 279, 281.

1327 UNITED STATES DEPARTMENT OF AGRICULTURE
Plant Diseases. Yearbook of Agriculture. 1953. USDA,
Washington, D.C. 1953.

1328 UNITED STATES DEPARTMENT OF AGRICULTURE
New inbred lines for millet hybrids. Agric. Res. (Wash.)
1965, 14: No. 4: p. 15.

1329 UNITED STATES DEPARTMENT OF AGRICULTURE
Economic Research Service. Review of 1965 and outlook

121   Uppal

for 1966. USDA, ERS. 152 and 153, 1966. Washington,
D.C.

1330 UPPAL, B.N. and DESAI, M.K.
Physiologic specialization in Sclerospora graminicola.
Phytopathology 21:337-338. 1931.

1331 UPPAL, B.N. and KAMAT, M.N.
Artificial infection of Pennisetum typhoideum by Sclerospora graminicola. Agric. J. India 23:309-310. 1928.

1332 USKACH, Kh. Ya and SYROEGIN, Yu V.
Chemical weeding of proso millet. Zemledelie 1966. 6:20-
21.

1333 USMAN, S.
A new insect pest of ragi in Mysore. Mysore J. Agric.
Sci. 1(4):256-258. Illus. 1967.

1334 UTTAMAN, P.
Parasitism of Striga lutea Lour. on rice and methods to
protect rice plant against Striga. Madras Agr. J. 37:99-
118. 1950.

1335 UTTAR PRADESH
Administrative Report of the Department of Agriculture,
Uttar Pradesh for year ending 30 June, 1949. 1950.

1336 VAAL, V. Der
Millet. Agric. Gaz. N.S.W. 7(1896) No. 3 pp. 132-135.
Exp. Sta. Rec. 8:125. 1896-1897.

1337 VAGER, R.M. and KARPENKO, G.A.
On the chemical composition of winter wheat mosaic virus.
Vop. Virusol. 7(4):106-109. 1962.

1338 VALLAEYS, G.
Coix lachryma-jobi. Bull. Agric. Congo Belge 1948, 39:
247-304.

1339 VAN BUUREEN, H., Jr.
Poona Agric. Coll. Mag. 5 and 6, 1915.

1340 VAN-DEN-BROEK, P.W.
The djalibras (Coix lachryma-jobi). Teysmannia 29(1):
59-61.

1341 VAN ROYEN, Wm.
Atlas of the World's Resources. The Agricultural Resources of the world. 1:78-82. 1954.

1342 VAN VAERENBERCH, R.
Le Coix lachryma-jobi en remplacement du mais jaune
dans l'engraissement du porc.   Bull. Agric. Congo 1961,
52:271-277.

1343 VARENICA, E. T.
Characteristics of some varieties of Setaria viridis.
Selek. Semenovdstvo 1950, 17: No. 6: 38-41.

1344 VARENICA, E. T.
Some results of the work on breeding and mechanization.
Selek. Semenovodstvo 1960: No. 1: 37-41.

1345 VASEY, H. E.
Millet smuts and their control.   Colo. Agric. Exp. Sta.
Bull. 242. 1918.

1346 VASIL'EV, P. P.
Emasculation of millet by hot air.   Selek. Semenovodstvo
1966, 31: No. 3: 73-77.

1347 VASIL'EVA, N. G.
The effect of high temperature and moisture of the soil
on the change in physiological characteristics of the water
regime.   (In: Biological basis of Agricultural Irrigation.)   Akad. Nauk. SSSR. Moscow 277-289. 1957.

1348 VASUDEVA, R. S.
Report of the Division of Mycology and Plant Pathology.
Sci. Rep. Indian Agric. Res. Inst. 1954-1955.   pp. 87-
101. 1957.

1349 VASUDEVA, R. S. and SURYANARAYANA, D.
The present position of research on millet diseases in India.   Proceedings of the Millet Workers Conference, Kolhapur, 1955.   Indian Council of Agricultural Research,
New Delhi, 1955.

1350 VAVILOV, N. I.
The origin, variation, immunity and breeding of cultivated plants.   The Ronald Press Co., New York. 1951.

1351 VAVILOVA, L. M.
Fifty years of work on breeding and seed production at
the Tulun Breeding Station (millet).   Irkutsk 1963: From
Referat. Zh. 1964: Abstr. 11, 55, 160.

1352 VEERASWAMY, R., ANAVARADHAM, L., and SREENI-
VASAN, V.
Optimum time of sowing for hybrid cumbu X-3 under irrigation in Satyamangalam tract.   Madras Agr. J. 1965,
52. No. 6, 286.

123   Vengris

1353 VENGRIS, J., HILL, E.R. and FIELD, D.L.
Clipping and regrowth of barnyard grass.   Crop Sci. 6
(4):342-344. 1966.

1354 VENKATANADHACHARY, G., HARIBABU, P. and GOPAL-
AM, B.
Improvement of grain yields in irrigated ragi (finger mil-
let).   Andhra Agric. J. 3(1):11-16. 1966.

1355 VENKATA-RAO, M.K.
Report of the work done in Mycology Section.   Report of
the Department of Agriculture, Mysore, 1928-1929.
1930.

1356 VENKATESWARLU, J. and PANTULU, J.V.
Morphology of pachytene chromosomes in pearl millet.
J. Hered. 1968. 59: No. 1: 69-70.

1357 VIDAL, P.
Influence of pedoclimatic factors on nutrition and production of Pennisetum millet cultivated in Senegal.   Paper
presented at the CCTA/FAO Symposium on Savannah Zone
Cereals. Dakar, Aug. 29 to 4 Sept. 1962.

1358 VIDAL, P.
The growth and mineral nutrition of millets (Pennisetum)
cultivated in Senegal.   Agrop. Tron. Paris 1963, 18:587-
668.

1359 VIEGAS, G.P.
Aspectos da cultura do ceral "adlay."   Bragantia 1951,
Nos. 1-3, 81 to 86.

1360 VIJAYAN, K.M. and NATARAJAN, S.
A note on the fungicidal control of ragi blast at Tindivan-
am. Madras Agr. J. 54(9):485-486. 1967.

1361 VIJAYARAGHAVAN, C.
Annual Report of the Millet Breeding Station, Coimbatore
for the year ending 1946-1947. pp. 24.

1362 VINALL, H.N.
Foxtail millet: its culture and utilization in the United
States. USDA Farmers Bull. 793. 1917.

1363 VINCENTE-CHANDLER, J., SILVA, S. and FIGARELLA, J.
Effects of nitrogen fertilization and frequency of cutting
on yield and composition of napier grass in Puerto Rico.
J. Agric. Univ. Puerto Rico 1959. 43:215-227.

1364 VLASOVA, N.I.
The biological peculiarities of smut in foxtail millet and
Hungarian grass. Byull. Vses. N.-I. Inst. Kukuruzy
1957(1):38-41. 1957.

1365 VOIGT, R. L.
Gahi-2 hybrid millet seed production in the Salt River Valley. Rep. Ariz. Agric. Exp. Sta. 1963, No. 215: pp. 7.

1366 VRESKY, F.
A brief description of some millet varieties and their uses in Czechoslovakian agriculture. Sb. Csl. Acad. Zemed. 1955: 28: 597-606.

1367 WALLACE, C. R.
The black beetle pest as it affects coastal dairy farmers. Agric. Gaz. N. S. W. 57(3):121-124. Sydney, 1946.

1368 WALLACE, G. B.
Plant Pathology. Rep. Dep. Agric. Tanganyika, 1948. pp. 45. 1950.

1369 WALLACE, G. B. and WALLACE, M.
Tanganyika Territory fungus list. Recent records XI. Mycol. Cir. Dep. Agric. Tanganyika 24, pp. 5. 1948.

1070 WALLACE, G. B. and WALLACE, M.
A second supplement to the revised list of plant diseases in Tanganyika Territory. East Afr. Agr. J. 13:61-64. 1947.

1371 WALSTER, H. L., HOOPER, T. H., MOOMAW, L., THOMPSON, O. A. and JORGENSON, L.
Field crop investigations in North Dakota 1923-1925. N. Dak. Sta. Bull., 194, pp. 9-23, 54, 80-82, 89. 1926.

1372 WANG, C. S.
Viability and longevity of chlamyoospores of Ustilago crameri. Phytopathology 26:1086-1087. 1936.

1373 WANG, C. S.
The formation of chlamydospores of Ustilago crameri on artificial media. Phytopathology 28:860-861. 1938.

1374 WANG, C. S.
Studies on the cytology of Ustilago crameri. Phytopathology 33:1122-1133. 1943.

1375 WATANABE, Yasushi and HIROKAWA, Humihiko
Auto Ecological Studies on the annual weeds in Tokachi. II. The effect of emergence on growth and development. Res. Bull. Hokkaido Nat. Agr. Exp. Sta. 93: p. 7-13. Illus. 1968. (Jap. and Eng. summ.).

1376 WATANABE, Yasushi and HIROKAWA, Humihiko
Auto ecological studies on the annual weeds in Tokachi.

III. The effect of time of emergence on the seed production. Res. Bull. Hokkaido Nat. Agr. Exp. Sta. 93: 16-
22. Illus. 1968.

1377 WATSON, K. A.
Fertilizers in Northern Nigeria. Current utilization and
recommendations for their utilization. Samaru Res. Bull.
38, 1964, pp. 20.

1378 WATSON, V. H., THURMAN, C. W. and WARD, C. Y.
Forage hybrids yield well in Station tests. Miss. Fm.
Res. 1967. 30. No. 4. 1 and 8 (Miss. Exp. Sta. State
College).

1379 WATT, G.
The commercial product of India. John Murray, London,
1908.

1380 WEATHERWAX, P.
Correlations between fruit covering and the embryo structure in some gramineae. Amer. J. Bot. 1961, 48: p.
537.

1381 WEBER, R. P. and SIMPSON, G. M.
Influence of water on wild rice (Zizania aquatica L.)
grown in a prairie soil. Can. J. Pl. Sci. 47(6):657-663.
1967.

1382 WELLS, H. D. and BURTON, G. W.
Helminthosporium setariae on pearl millet, Pennisetum
typhoides as affected by age of host and host differences.
Crop Sci. 7(6):621-622, 1967.

1383 WELLS, H. D.
Effectiveness of two 1, 4 oxathiin derivatives for control
of Tolyposporium smut of pearl millet. Pl. Dis. Reptr.
51(6):468-469. 1967.

1384 WELLS, H. D.
Effects of temperature on pathogenicity of Helminthosporium setariae on seedlings of pearl millet, Pennisetum typhoides. Phytopathology 57(9): 1002. 1967.

1385 WELLS, H. D. and WINSTEAD, E. E.
Seed borne fungi in Georgia. Grown and western grown
pearl millet seed on sale in Georgia during 1960. Pl.
Dis. Reptr. 49:487-489. 1965.

1386 WELTON, F. A.
Sorghum and millet. Mo. Bull. Ohio Sta. 1(6):168-174.
1916. Exp. Sta. Rec. 35:529.

1387 WEN, Chin-Chen
A study on the millet stem fly, Atherigona biseta Karl.
Acta Phytophylac. Sin. 3(1):41-47. 1964. From: Sci.
Abstr. China Biol. Sci. No. 740. 1965.

1388 WESTER, P.J.
The food plants of the Philippines. Philipp. Agric. Rev.
16(3):221-222. 1921.

1389 WESTER, P.J.
Notes on Adlay. Philipp. Agric. Rev. 13(3):217-222. 1920.

1390 WESTER, P.J.
Adlay, a new grain from the Orient. J. Hered. 13(5):
221-227. 1922.

1391 WESTON, W.H. and UPPAL, B.N.
The basis for Sclerospora sorghi as a species. Phytopathology 22:579-586. 1932.

1392 WESTON, W.H. and WEBER, G.F.
Downy mildew (Sclerospora graminicola) on everglade mil-
let in Florida. J. Agr. Res. 36(11):935-967. 1928.
Exp. Sta. Rec. 60:51.

1393 WHEATLEY, P.E. and CROWE, T.J.
"Pest Handbook." Kenya Dept. of Agric. Nairobi, 1967.

1394 WHEELER, C.F.
Plant disease, Mich. Sta. Rpt. 1897, pp. 99, 100. Exp.
Sta. Rec. 11:58. 1899-1900.

1395 WHEELER, W.A. and HILL, D.D.
A revised list of plant diseases in Rhodesia. Kirkia 5
(2):87-196. 1966.

1396 WHITESIDE, J.O.
A revised list of plant diseases in Rhodesia. Kirkia 5(2):
87-196. 1966.

1397 WIEHE, P.O.
The plant diseases of Nyasaland. Mycol. Pap. 50. Commonwealth Mycology Institute, Kew. 1953.

1398 WIEHE, P.O.
The Plant Diseases of Nyasaland. Mycol Pap. 53. Commonwealth Mycological Inst. Kew. 1950.

1399 WIESEROWA, Aleksandra
Wczesnosredniowieczne szczatki zboz i chwastow z prze-
mysla (Early medieval remains of cereals and weeds
from Przemysl Southeast Poland). Folia Quarternaria
28:1-16. 1967.

1400 WILLIAMS, T. A.
      Millets. USDA Yearbook 1898, pp. 267-290. 1898.

1401 WILSON, J. W. and SKINNER, H. G.
      Millet for fattening swine. S. Dak. Sta. Bull. 83, 1904-
      1905. Exp. Sta. Rec. 16: 294.

1402 WILSON, J. W. and SKINNER, H. G.
      Speltz and millet for the production of baby beef. S.
      Dak. Sta. Bull. 97, pp. 61-74. 1907. Exp. Sta. Rec. 18:
      261.

1403 WILSON, R. G.
      Machine sown pastures on the Darling Downs. Qd. Agric.
      J. 82(2):63-70 and (3):125-135. 1956.

1404 WINTON, A. L. and WINTON, K. B.
      The structure and composition of foods. Vol. I: Cereals,
      Starch, Oil Seeds, Nuts, Forage Plants. John Wiley and
      Sons, Inc., New York. 1932.

1405 WISNIEWSKI, P.
      Influence of acetic acid on control of millet smut (U. pani-
      cimiliacei) and on the germination of millet grains. Pol-
      ish Agric. & Forest Ann. 22:363-378. 1929.

1406 WITSCH, H. Von
      The effect of the length of day on the content of colouring
      matter and aneurin in the leaves of long and short day
      plants. Z. Bot. 47(1-2):121-144. 1959.

1407 WITTE, H.
      An experiment with corn and other North American soil-
      ing and silage crops at Svalof in 1920. Sveriges Utsades-
      for. Tidskr. 33(5):268-275. 1923. Exp. Sta. Rec. 50:
      735.

1408 WOOD, R.
      A notebook on agricultural facts and figures. pp. 194.
      Supt. Govt. Press, Madras, 1920.

1409 WORLD FARMING
      Break through for millet (Pennisetum typhoides). Wld.
      Fmg. 1968. 10: No. 11, p. 37.

1410 WRIGHT, W. H.
      Millets (Panicum, Setaria and Pennisetum) in Handbook on
      seed testing. Assoc. Off. Seed Analysts. pp. 1-4. 1941.

1411 WU, S. -H. and TSAI, C. K.
      Cytological observations on the F_2 hybrid rice (Oryza
      sativa L.) x Pennisetum sp. Chi-wu-hsueh PAO (J. Bot.)
      (Acta Bot. Sin. 1963, 11:293-307.)

1412 YABUNO, T.
Land and crops of Nepal Himalaya. Scientific results of
the Japanese expeditions to Nepal Himalaya 1952-53. Vol.
II. Echinochloa pp. 255-259. Fauna and Flora Res. Soc.
Kyoto Japan, 1956, pp. x, 529.

1413 YABUNO, T.
Oryza sativa and Echinochloa crusgalli var. oryzicola
Ohwi. Rep. Kihara Inst. Biol. Res. 1961, No. 12:29-34.

1414 YABUNO, T.
A hybrid between the tetraploid annual Echinochloa crusgalli var. oryzicola and 60-1, a tetraploid perennial strain
from West Bengal. Rep. Kihara Inst. Biol. Res. 1962,
No. 13:52-56.

1415 YABUNO, Tomosaburo
Biosystematic study of the genus Echinochloa (gramineae).
Jap. J. Bot. 19(2):277-323. 1966.

1416 YAKIMENKO, A. F.
Sowing times and methods for millet and buckwheat. Zem
ledelie 1965, No. 5:59-60.

1417 YAMURA, A.
Genetical studies in the genus Coix. Jap. J. Genet.
1949, 24:180-182.

1418 YANG, M. H.
Preliminary observations on experiments on hybridizing
rice and Kweichou Echinochloa. Nung-yeh-nsueh Pao (J.
Agric.) Acta Agr. Sin. 1960, 11:77-82.

1419 YAROSH, N. R.
The content and composition of protein and starch in grain
of proso millet of various ecological and geographical
groups. Tr. Prik. Bot. Genet. Selek. 1965, 37: No. 1:
50-58.

1420 YEFIMOV, I. T.
The growth and development of millet when sown on stubble and irrigated. Agrobiologiya 1956(3):110-112. 1956.

1421 YOUNGMAN, W.
Annual Report Economic Botany Department. Department
of Agriculture Central Provinces and Bihar, 1924-25.

1422 YOUNGMAN, W. and ROY, S. C.
Pollination methods among the lesser millets. Agric. J.
India 18:580-583. 1923.

1423 YOUNGNER, V. B.
Low temperature induced male sterility in male-fertile
Pennisetum clandestinum.   Science 1961, 133:577-578.

1424 YU, T. F.
Further studies on kernel smut resistance in millet.
Chin. J. Exp. Biol. 1:235-240.   (Cf. Rev. Appl. Mycol.
16:741) 1937.

1425 YU, T. F.
Inheritance of kernel smut resistance in millet crosses.
Sci. Rec. Acad. Sin. 1942, 1 Nos. 1 and 2: 248-250.

1426 YU, T. F.
Reaction of improved millet varieties to infection with
downy mildew (Sclerospora graminicola Schroet.)   Chin.
J. Sci. Agr. 1944, 1:199-203.

1427 ZAIKINA, I. N.
On methods of obtaining large-grained forms of millet.
pp. 164-172.

1428 ZAKHARCHISHINA, V. A. and PHIPENKO, T. I.
The uptake of mineral elements by buckwheat, oats and
millet and their translocation to various plant parts.
Nauch. Dokl. Vyssh. Shk., Biol. Nauk.   7(43):86-89.
1967.

1429 ZAKLADNII, G.
An experiment on rapid reproduction of selected varieties
of millet in collective farms.   Opyt. Agron. No. 4, 56-
68.   1941.

1430 ZAKLADNYJ, G. A.
Some problems of directed heritable changes in millet.
Vestn. Sel'-Kh. Nauk. 1963, No. 3:119-120.

1431 ZAKLADNYJ, G. A.
Some questions concerning directed change of inheritance
in millet.   Agrobiologiya 1953, 459-460.

1432 ZAVITZ, C. A.
Variety and other tests of field crops.   Ann. Rep. Ont.
Agr. Coll. Expt. Farm, 35:166-230.   1909.   Exp. Sta.
Rec. 23:332.

1433 ZAZHURILO, V. K. and SITNIKOVA, G. M.
Inter-relation between the virus of mosaic disease of
winter wheat and its vector (Deltocephalus striatus L.)
Proc. Lenin Acad. Agric. Sci. 6(11):27-29.   Moscow,
1941.

1434 ZDANOVA, L. P.
Comparative analysis of photoperiodic induction in short
and long day plants. Tr. Inst. Fiziol. Rast. im. K. A.
Timirjazeva 6:69-84. 1948.

1435 ZEBRAK, A. R. and AFANASJEVA, A. S.
Relative fertility of tetraploid and diploid Panicum millets
under field conditions. Dokl. Akad. Nauk. SSSR 1948,
61:525-528.

1436 ZHANG, C. F.
Studies on the systemic infection induced by sporangia of
Sclerospora graminicola in millet. Acta Phytophylac. Sin.
4(2):163-168. 1965.

1437 ZHEBRAK, E. A., RUNKOVA, L. V. and NIKOL'SKII, Yu K.
Growth substances in dry seeds of diploid and tetraploid
millet. 13v. Akad. Nauk. Belaruss. S. S. R. (Ser. Biol.
Nauk 1967. No. 4 109-111 (Bibl. 9 Ru).

1438 ZIMMERMANN, K.
Hirse als kornerfrucht (Millet as a grain crop). Dtsch.
Landw. 1961, 0 101 100.

1439 ZUMATOV, A. Z.
Micurinist millet growers of Kazahstan. Zemledelie
1955: No. 10:62-64.

AUTHOR INDEX
(with language indicators--see page 202)

131

SUBJECT INDEX

Each of the following species indexes is arranged under a number of general headings. Not all the headings appear in each species index (and a few differ from those below), but the overall structure and order of headings within a species are as follows:

General and historical
Anatomy, morphology
    and histology
Physiology and biochemistry
Taxonomy
Genetics
Cytogenetics and cytology
Breeding and varieties
Agronomy
Climate
Cropping systems
Fertilisers
Irrigation

Sowing and transplanting
Weeds and control
Diseases
Pests
Composition and nutritive value
Fodder
Grain
Food (or Food uses)
Extract
Milling
Oil
Storage
Utilization

MILLETS (General)

GENERAL AND HISTORICAL

Atlas of World's Resources   1341E
Crops of the Bombay Presidency   964E
Economic products of Malay Peninsula   165E
Experimental station records 1899-1946 U.S.A.   1325E
Field crops of India   526E
Field and garden crops of the Bombay Presidency   374E
Field and garden crops of the Punjab   811E
Geography and History of Millets   636E
Handbook of Agriculture   507E
Indian Medicinal Plants   603E
Industries of Russia--Agriculture and Forestry   275E
Manchurian millets   469E
Millet   42E, 54Rs, 116Ge, 243E, 431E, 1124E, 1128E,
    1138E, 1150E, 1276E, 1336E 1386E, 1400E
Millets from Bronze Age   1310It

TAXONOMY   445Ge, 490Ch, 617Ge

Agrobotanical research--Bulgaria   490Bu
Classification and distribution South Asia   620E
Flora of: Presidency of Bombay   269E
---: British India   483E
---: Presidency of Madras   373E
---: Tropical Africa   1242E
Grasses and Pastures of S. Africa   801E
Systematic collection--India   1032E
Text book of Punjab agriculture   1107E

GENETICS

Chromosome numbers   169, 637, 639, 1093, 1196E
Mutation experiments   325Ge
Polyploidy   837Rs, 937J
Quantitative inheritance   180E
Right and left-handedness   258E
X-irradiation induced mutagens   132Hu

CYTOGENETICS AND CYTOLOGY   169, 630, 1199E
Colchicine induced autotetraploidy   9Rs
Directed heritable changes   1430Rs
Taiwan grasses   233Ch

BREEDING AND VARIETIES   442E, 642Ge, 682E, 815E, 939Rs,
                         983Rs, 1115Rs, 1264Rs, 1265Rs,
                         1305Ge, 1306Ge, 1350E
Canada   803Fr
Czechoslovakia   1366Cz
East Africa   472E
Germany   1305, 1306Ge
India   124, 504, 726, 1067, 1085, 1238E
New Zealand   1276E
Niger   218Fr
Senegal   125Fr
U.S.A.   808, 1371E
U.S.S.R.   298, 426, 442, 496, 532, 772, 857, 899, 995,
           996, 1344, 1351, 1439Rs
Breeding for: drought resistance   926Rs
---: larger grain   1427E
Change of heritable basis   344Rs
Diallel crosses   427, 428, 468, 548, 618E
Drought adaptation   315E
Emasculation by hot air   542, 1346Rs
Emasculation, thermal   1020Rs
Hybridisation   541Rs
        vegetative   408, 626Rs
Inbred lines for hybrids   1328E
Methods   410Rs, 830Rs, 1088E
Millet Improvement Programme--India   508, 509, 510, 511,
        512E

BREEDING AND VARIETIES (cont.)
    Novelties for the trade   387Ge
    Reproduction of selected varieties   1429Rs
    Seeds--quantity of dust required   981Ge
    Selection and seed production   235Rs
    Selfing effect on yield   1183Ch
    Smut immunity   699Rs
    Varieties   306Rs, 413E, 566Rs, 567Rs, 604Rs, 640Rs, 641Ge,
        657Rs, 772Rs, 775Rs, 804Rs, 805Rs, 986Rs, 987Rs
        Kamysin   123   616Rs
        Mironovka 85   901Rs
        Orenburg 42   307Rs
        Ta-Hung P'ao   490Ch
        Veselyj-podol 38   616Rs
    Variety trials   57E, 366E

AGRONOMY

    Canada   1144, 1145, 1146, 1432E
    India   257, 841, 1421E
        increasing food production   1086E
        irrigated farming   814E
        water requirements   670, 1041E
    Malawi   030, 000E
    Morocco   911Fr
    Poland   679Pl
    Senegal   1000Fr
    Sudan   1312E
    Tanzania   694, 943E
    Uganda   953, 1318E
    U.S.A.   214, 808, 1072, 1167, 1225E
        supplementary and emergency crops   56E
    U.S.S.R.   250, 1141, 1222, 1324Rs
        semi-desert steppe reclamation   760, 1123Rs
        utilization of: Don area sands   331Rs
        ---: water meadows   862Rs

CLIMATE

    Ecological crop geography--agro-climatic analogues   928E
    Effect on growth and development   212E
    Evapotranspiration--climate relations   449E
    Rainfall and yield   894E

CROPPING SYSTEMS

    As a second crop   539Rs
    Cover crop for orchards   345E
    Fallowing   971E
    Mixed cropping   44, 77, 873E
    Principles of field crop production   773E
    Rotation   893Rs
    --- and soil fertility   324E

FERTILISERS

Chemical treatment   1217Rs
Effect on germination   828Bu
Experiments   321Rs, 1187Rs
India   1087E
Magnesium   784E
Nitrogen uptake   43Rs
Phosphate absorption by excised roots   921E
Productivity and fertiliser composition   359Rs
Response   462, 802E
--- to phosphate   683Ch
Super phosphate   7Rs
Zinc   8Rs

IRRIGATION

Accelerates smut removal   358Rs
Kenya   108E
Maturation and harvesting   531Rs
Mechanisation   1344Rs
Plant nutrition: nutritive solution   613Rs
---: soil phosphorus   946Rs
Seed analysis   753Rs
Soil--effect of drying   127E
Soil--trace of elements   434Ge
Water requirements   316E

SOWING

Autumn sowing   1172Rs
Planting depths   1103Ge
Stubble sowing and irrigation   1420Rs
Time and methods   1416Rs
Wide spacing   700Rs

WEEDS AND CONTROL

Control in legumes and pastures   712E
Control--chemical   1182Rs
---: sowing   31Rs
Foxtail giant   1162E
Foxtail giant and yellow   1140E
Response to herbicides   1262E
Sensitivity of sodium salt of 2, 4-D   1157Rs
Spraying trials   29Ge, 30Ge
Swainsonpea   1108E

DISEASES   204E, 280Rs, 314E, 474J, 520E, 759Rs, 1050E, 1213E,
1219E, 1246E, 1269E, 1327E, 1394E

Australia   710E, 1185E
Canada   277E

Stored grain pests 1025E
Use of Aldrin 596Rs

Bird Pests

Queleo spp. 1018E

FODDER 1315E

Arid zones 519It, 658It
Australia 146E
India 954, 955E
Sweden 1407Sw
U. S. A. 115, 381, 486, 491, 605, 1147, 1184E

Chemical Composition and Nutritive Value 245E, 378Rs, 945E

Amino acid content 698E
Digestibility 666, 905E
Feed for: baby beef 1402E
---: horses 473E
---: swine 1401E
Hay 660, 665E
Hydrocyanic acid 156, 369E
Straw 378Rs, 956E
    processing 770E
Toxicity to sheep 107, 1245Rs

GRAIN 1315E, 1438Ge

U. S. A. 381, 491, 1184E
    Characteristics 690Ge
    Competition 540E
    Drying 337Rs

Chemical Composition and Nutritive Value

Amino acids 1047Rs
Boza--microbiology of 26E
Fat soluble vitamins 1243E
Feeding value 202Fr, 462E, 1287Rs
---for pigs 836E
---for turkeys 411E
Flour 1031It
Hulls 332Rs
Local food--Uganda 164E
Manganese content 840E
Protein values 93E, 688Ch
--- and starch 538Rs
Selenium content 447E
Starch--hydrolysis by diastase 1292J
Tryptophan 350Rs

FOOD 5Fr

Baking 1194Po, 1193It
Gruel crops 388Rs
Indian diets 73E
Pre-Colombian Mexican Indians 208E
Structure and composition 1404E
Taste quality 671Fr

EXTRACT

Diastatic properties 607Rs
Production 224E

MILLING 4, 6, 117Fr

OIL

Chemical composition 112Rs
Hydrogenation 301Rs
Vitamin composition 113Rs

STORAGE 364, 632E

UTILIZATION

India 437, 726, 750, 1379E
Morocco 911Fr
Nigeria 365E
U.S.S.R. 79, 80Rs

BRACHIARIA RANOSUM

PHYSIOLOGY
Dormancy 35E
Seed treatment and germination 36E

CYTOGENETICS
Supernumerary Chromosomes 1196E

AGRONOMY 1278E
India 234E
U.S.A. 273E
Soil phosphorus 973E
Stress--insecticides 92E

DISEASES
Maize mosaic virus 1348E

PESTS
armyworms 129E

FODDER
    For steers 98E
    Management 968E

FOOD USES
    Philippines 1388E

DIGITARIA EXILIS

ORIGIN 238Fr

AGRONOMY
    Africa and Europe--culture 1001Fr
    Gambia--production 372E
    Nigeria--fertilizer use 1377E

CHEMICAL COMPOSITION 211Fr

DIGITARIA IBURIA

ORIGIN 238Fr

AGRONOMY
    West Africa 998Fr

NUTRITIVE VALUE 667Fr

DIGITARIA

MONOGRAPH 471Du

COIX LACHRYMA-JOBI

THE PLANT
    Brazil--origin and distribution 1158E
    Burma, Ceylon, India, Pakistan 139E
    South Africa 801E

BOTANY
    Morphogenesis of mesocotyls and internodes 584E
    Morphology 527Fr, 812Fr, 1380E

GENETICS
    Inheritance of characters 1417J
    Waxy endosperm 583E

CYTOGENETICS AND CYTOLOGY 503E
    Intergeneric crosses with Zea 453J

BREEDING 851J, 852J, 853J, 854J, 855J, 856J, 857J, 936Fr
    Belgian Congo 260, 261Fr

BREEDING (cont.)
     New Guinea   152Fr

DISEASES
     Helminthosporium turcicum, H. maydis and H. carbonum
          1106E
     Host range of bacterial wilt of sweet corn   343E
     Hosts of Helminthosporium, India   230E
     Leaf blotch   217It
     Phyllosticta   95Sp
     Puccinia sorghi   758E
     Smut   474J, 1300E
     Ustilago coices   247E

PESTS
     Stem borers   495E

THE PRODUCT
     Milling tests and equipment   466E
     Nutritive value   296Sp, 305Sp, 529Sp
     ---for: chickens   1042Sp
     ---: human food   1310E
     ---: pigs   1156, 1342Fr

UTILIZATION   783E
     Belgian Congo   1338Fr
     Brazil   1359Sp
     Dutch East Indies   611Du
     Philippines   978, 982, 1388, 1389, 1390E

COIX AQUATICA

Chiasma frequency   1251E

Helminthosporium   230E

Heterochromatin and non-homologous associations   621E

COIX Sp.

Botany of Asiatic types   911, 912J

Cytology and Morphology of new species   624E

Interspecific hybrids   623E

Nucleolus   622E

ECHINOCHLOA CRUSGALLI-PANICUM CRUSGALLI

ORIGIN AND CULTIVATION   1400E
    India   131E
    U.S.A.   478E
    U.S.S.R.   532Rs

MORPHOLOGY AND HISTOLOGY   801E, 1404E

PHYSIOLOGY AND BIOCHEMISTRY
    Germination under irrigated and non-irrigated conditions
        577J
    Growth: effect of pH level   382E
    ---: effect of soil reaction   1274Pl
    Seed dormancy   241, 242J

GENETICS, CYTOGENETICS AND CYTOLOGY
    Base exchange of roots   368J
    Biosystematic study   1415J
    Chromosome numbers   1412J
    Genecological studies   872J

BREEDING
    Improved varieties   1195E

AGRONOMY
    North America   58E
    Philippines   549E

CROPPING SYSTEMS
    Protection for groundnut crop   24E
    Rotations   1283J

SOWING
    Effect of soil depth on emergence   292E
    Methods   984Rs

WEEDS
    Control: with auretryne   709E
    ---: in lucerne   712E
    ---: in maize and field beans   708Rs
    ---: in rice field   15E, 576E, 693E, 833E, 872E, 1220E
    Pre-emergence herbicides   556E
    Rice fields   339J, 606E

DISEASES
    Cacoecia epicyrata Meyrick   1030E
    Helminthosporium   917J
    Hoja Blanca   423E
    Infection by H. monocerae   563Yg
    Sorosporium bullatum   271E

DISEASES (cont.)
    Sphacelotheca destruens, cross inoculation on maize   1185Rs
    Ustilago paradoxa   646E, 1180E

PESTS
    Heterococcus nigeriensis   456E
    Leptocorisa varicoruis F   1169E

FODDER
    Australia   15E, 327E, 904E, 1403E
    U. S. A.   346E, 1325E
    U. S. S. R.   532E, 863Rs
    Clipping and fodder production   1354E
    Nutritive value   685E, 756Pl, 1079E

GRAIN
    Nutritive value   625J, 1404E
    Oil   340Sp
    Silage   45E
    Wildlife food   19E

UTILIZATION
    Hawaii   1132E
    India   882E

ECHINOCHLOA COLONA

ECOLOGY
    India   153E
    U. S. A.   478E

DISEASES
    Hoja blanca   370E, 423E
    Nematode induced galls   128E
    Sogata cubana, a virus vector   483Du
    Sogata orizicola   371Sp
        E. crusgalli var. orizicola   1413J
        interspecific hybrid   1414J
        seed dormancy   49J
        E. crusgalli var. edulisa
        fodder in Australia   590E
        E. macrocarpo
        entomological control   596Rs
        Kweichou Echinochloa
        intergenetic crosses   1418Ch
        E. crusgalli var. praticola   1375J, 1376J

ELEUSINE CORACANA

MORPHOLOGY   898, 1319E
    Effect of thiamine on root growth   1143E

PHYSIOLOGY AND BIOCHEMISTRY
    Effect of salinity   571E
    Effect of seed soaking   293E
    Inhibitor in "seed-coat"   1142E
    Seed treatments   226, 815E
    Water requirements   1041E

GENETICS, CYTOGENETICS AND CYTOLOGY
    Dicentric chromosomes   1198E
    Genetic variability   582, 884, 885E

BREEDING   1282E
    Discriminant function for selection for yield   742E
    Early maturing variety   892E
    Emasculation by hot water technique   1043E
    India   906, 947E
    New species   317E
    Strain Co 8.   728, 730, 1170E
    Variety trials   1134E

AGRONOMY   493, 1160E
    Belgian Congo   268Fr
    Ethiopia   294E
    India   256, 750, 888E
    Kenya   1148E
    Malawi   304, 931E
    Malaya   1232E
    Philippines   1223E
    Rhodesia   551E (Zambia 1097, 1098E)
    Tanzania   694E
    Uganda   334, 452E

FERTILISERS
    Azotobacter   1089E
    Experiments   869, 1090, 1244, 1275, 1325E
    Foliar nutrient sprays   81E
    Indicator plants   1299E
    Irrigation   1084E, 1352E
    Micronutrients   414E
    Nitrogen fixation in latosolic soil under glass   834E
    Nitrogen levels and tillering 630E
    Nitrogen and yield potential   575E
    Nutrient uptake 1127E
    Nutritive content and fertiliser levels 631E
    Optimum dose   34E
    Plant type and harvest index   1239E

FERTILISERS (cont.)
    Response to: calcium ammonium nitrate   516E
    ---: potassium phosphate   497E
    ---: zinc   1078E
    Yield--effect of fertilisers and planning density   420E

SOWING AND TRANSPLANTING
    Clonal propagation   748E
    High altitudes   157Fr
    Sowing rates   268Fr
    Sowing time and plant population   1082E
    Sowing times and transplanting   967, 1135E
    Transplanting a short season crop   747E
    Transplanting seedlings   318E
    Storage in underground pits   889E
    Stored seed treatment with fungicides   226E

DISEASES   789E, 914J
    Bacterial blight   309, 310E
    Bacterial leaf spot--India   1077E
    Colletotrichum gramincolum   248E
    Control of ragi blast by fungicides   1046, 1360E
    Epiphytotic disease   416E
    Helminthosporium nodulosum and virus   559, 816E
    H. rostratum   675E
    Malawi   1398E
    Mosaic virus   1080E
    Ozorium wilt   221E
    Reaction to Pellicularia Rolfsii and Nematode   415E
    Seedborne fungi--distribution and control   424E
    Smut   647E
    Tanzania   1368, 1369, 1370E
    Uganda   1214E
    Xanthomonas and fusarium persistence in soil   1075E

PESTS   1254E, 1387Ch
    Azazia rubicans   1029E
    Cacoecia epicyrta Meyrick   1030E
    Colemania sphenaroides   1252E
    Holotrichia sp. and control   1074E
    Leptocorisa varicornis   1169E
    Locusts in Uganda   454E
    Lygus simonyi   1296E
    Sitotroga cerealella   1255E
    Stored grain pests in East and Central Africa   944E
    Uganda pests   290, 515E

COMPOSITION AND NUTRITIVE VALUE   552, 581, 1065, 1404E
    Amino acid   86E, 1279J
    Calcium and phosphorus   403E
    Hydrocyanic acid   156E
    Indian foodstuffs   287, 288, 289, 312E
    Protein   751, 896E

AGRONOMY (cont.)
       water use   338
    Ethiopia
       culture   249It
       field trials   294E
    Southern Rhodesia
       rotation crops   844E

DISEASES
    Anthracnose   215It
    Fungal diseases   216It
    Leaf blight   918J
    Uromyces eragrostidis   419It

CHEMICAL COMPOSITION   537E, 597Sw, 797E, 1257E
    of hay   412E
    of oil   659It

ERAGROSTIS CURVULA

FERTILISER TRIALS   1297E

PANICUM MILIACEUM

GENERAL AND HISTORICAL   109Rs, 460E, 717E
    Early mediaeval remains   1399Pl
    Iron age plants   1163Ge
    Remains from Roman times   938Pl

ANATOMY, MORPHOLOGY AND HISTOLOGY
    Coloration of glumes   52Rs
    Morphology of inflorescence and flowers   664Ch

PHYSIOLOGY AND BIOCHEMISTRY
    Gibberellins--dynamics of growth substances   691Rs, 1437Rs
    Photosynthetic activity and crop structure   649Rs
    Rhizosphere bacteria   429Rs
    Seeds--germinability   705Rs
    Seeds--longevity   704Rs

TAXONOMY   53, 110, 707, 1209Rs

GENETICS--tetraploid and diploid   1435Rs

CYTOGENETICS AND CYTOLOGY   543E
    Accessory chromosomes   546E
    Interspecific relationships in phasic development   702Rs

BREEDING AND VARIETIES 138E, 209Fr, 545E, 785Rs, 1024Rs,
                                     1131Rs, 1154Rs, 1165Rs, 1188Rs
    Behaviour of populations and selections 88Sp
    Evaluation of hybrid progeny 703Rs
    Panicum millet 242Ch
    P. dewinterii (P. erectum 271E) 41E
    P. deustum 1096E
    Seed testing 1410E
    Smut resistance 706Rs
    Zn--early maturing 524Rs

AGRONOMY
    Colorado 145E
    Germany 579Ge
    Kenya 1148
    Philippines 549E
    U.S.S.R. 612Rs

    Agrobiological Classification 701Rs

CLIMATE AND WEATHER
    Agroclimatic indices for yield 1151Rs
    Meteorological factors and yield 1027Rs

CROPPING SYSTEMS
    Catch crop under irrigation 900E
    Rotations 1000 Fr
    Short season replacement crop 418E

FERTILISERS 572Rs, 678Rs
    Mineral uptake and yield 1216Cz
    Weed control 656Rs, 1332Rs

DISEASES 644Ge
    Bacterial strip 341E
    Fungus 1248E
    Pyriculariosis 1002Rs
    Rust 279E
    S. destruens and U. crameri 643Ge
    Smut 213E, 644Ge
        control by acetic acid 1405Pl
        inoculation 59Rs
        resistance 522Rs, 910Rs, 1152Pl, 1153Pl
        seed treatment 628Pl, 765Rs
    Virus--wheat streak mosaic 1433Rs
    Virus--inoculation tests with Ph. stewartii and Ph.
        rasculora 523Rs

PESTS
        Insect injury 655J
        Leptocorisa varicornis 1169E
        Stenodiplosis panici 941, 1028Rs

CHEMICAL COMPOSITION and Nutritive Value 113Rs, 1419E
        Carotenoids 1046R
        Digestibility 1221E
        Table food 450E

FODDER 206Sp

GRAIN 717E

FOOD
        Lambs 835E
        Lysine deficient for pigs 349E
        Swine 465E

UTILIZATION
        Food uses--Morocco 383Fr

PASPALUM SCROBICULATUM

AGRONOMY
        Ratoon cropping India 319E
        Response to nitrogen 762E

DISEASES
        Claviceps paspali 359E, 554E
        Ephelis japonica--Sierra Leone 297E
        Xanthomoras rubisorghi--India 1076E

STRAW--Nutritive Value 957E

Nutritive Value and toxicity of fungus-free grains 12E

UTILIZATION
        India 725E
        Malaya 1124E
        Medicinal use 303E

PASPALUM Sp.

Cytology 731E

Tropical American rusts 279E

Utilization--India 272E

PENNISETUM TYPHOIDES

ANATOMY, MORPHOLOGY AND HISTOLOGY   458E

    Anthesis 83E, 1160Fr
    Growth habit 76E
    Morphological and cytological response to mutagens   194E
    Morphology 89E, 405E
        of bristles 897E
        of pachytene chromosomes 1356E
        of pachytene chromosomes of P. purpureum   950E

PHYSIOLOGY AND BIOCHEMISTRY

    Base exchange of roots 368J
    Biology of flowering 600Rs
    Daylength response in Puerto Rico 90E
    Detrimental effect of selenium 970E
    Dormancy 35E
    Drought
        plant water relations 661E, 662E
        heat and drought stress 1260E
        resistance 356E
    Dynamics of natural gibberellins 691Rs
    Flooding tolerance 561E
    Germination 286, 816E
    Gibberellin--effect on stamen elongation 1159Fr
    Leaf efficiency 101E
    Metabolism 104E
    Moisture relations at permanent wilting stage 10E
    Moisture supply--effect on growth 875E
    Photosynthesis in low latitude environment 102E
    Photosynthetic productivity and growth in arid areas   861Rs
    Pollen storage and time of pollination 270E
    Pollination 93E
        by insects 676E
    Potassium assimilation 788E
    Reaction to cold and drought 602Rs
    Response to inundation 291E
    Root uptake of phosphorus 932E
    Salt tolerance of germination 1E
    Seasonal influence on water requirements and growth   404E
    Seed treatment with B-nine 815E
    Seed treatment with succinic acid 766, 767, 1202E
    Soil moisture and evapotranspiration 530E
    Water requirements: arid areas 862Rs
    ---: India 1041E
    ---: Lower Congo 335Fr

TAXONOMY   87E

GENETICS

CYTOGENETICS AND CYTOLOGY

Bajra-rapier grass hybrid 693E
Chiasma frequency 745E
Chromosome: accessory 1007, 1008, 1012E
---: karyormorphology 400E
---: miniature centric fragment 1014E
---: numbers 544, 1272E
---: variants 1003, 1004, 1006, 1009, 1010E
Colchicine induced tetraploidy 401E
Cytomorphology of allopolyploid progeny 1234E
Interspecific crosses 394, 500, 594, 962E
Interspecific hybrid with oryza 1411Ch
Interspecific hybrids 362, 800, 877, 1013, 1062, 1063,
  1194E
Interspecific hybridisation effect on seeds 396E
Karyotypes 1016E
Meiosis 949, 1069E
Mutagens: effect 1011E
---: response to 195E
Pennisetum, panicum 1096E
P. ciliare 1289E
P. ruppellii Steud 1061E
P. squamulatum 960E
Sterility--Induced with dimethyl arsinic acid 1015E

BREEDING AND VARIETIES 17E, 62E, 64E, 173E, 177E, 178E,
                        198E, 502E, 965E, 1068Fr, 1409E

Aims and methods 601Rs
  Australia 806E
  Chad 119Fr
  Congo 263Fr, 264Fr
  India 272, 867, 1034, 1038, 1361E
  Italy 1095It
  Morocco 974Fr
  Niger 870, 871Fr
  Nigeria 2E
  Philippines 549E
  Senegal 137, 352, 353, 1170Fr
  Uganda and East Africa 976E
  U.S.S.R. 785, 804, 805Rs
  Zambia (Northern Rhodesia) 1099E
  Bajra bristled S. 531 68
  Bajra hybrid HB-1 61, 70, 441, 729, 1090, 1139,
    1204E
  ---: T.55 832E
  ---: X3 727E
  ---: New 506E
  ---: White 1205E
  Breeding behavior and time of pollination 172E
  Commercial use of an interspecific hybrid 1013E
  Diallel cross analysis 738E
  Discriminant functions in selection 746E

Saudi Arabia 1229E
Senegal 1170, 1357, 1358Fr
    ecology of use 141Fr
South Africa 1298E
Swaziland 1273E
Tanzania 155E
Tropical Africa--Ecology 999Fr
Uganda 1299E
U.S.A. 362, 1261E
West Africa
    Mechanical cultivation 1313Fr
    Agroclimatological study 254E

CROPPING SYSTEMS AND ROTATIONS

Cover crop 933E
Cultural methods 907Fr, 980E
Field plot techniques 570E
Groundnuts intercropped with cereals 1161E
Mulching 839Fr
Rotations 300Fr, 653E, 903E, 979E, 1256Fr

FERTILISERS

Effect on: hybrid bajra (HB-1) 1090E
---: mineral content 1175E
---: Napier grass 1363E
---: vitamin content 711E
---: yield and quality 1206E
Green manuring 467E
    with acacia 227Fr
    of rice 437Fr
Influence of copper fertilization on copper content 118Af
Phosphate 589E
Phosphorus levels 1101E
Response: for hay 92E
---: of bajri-tur mixed crop 121E
---: to nitrogen 817, 859E
Soil fertility, regeneration of 300Fr
Trials 653, 771, 1164, 1273E
Urea 599E
    effect on grain 595E
Use in Nigeria 1377E
    Rhodesia and Malawi 1097E

PLANT NUTRITION

Effect of: sodium salts on growth and composition 302E
---: soil nutrients and pH on nitrate nitrogen and growth
    1218E
Micronutrients 820E
Nutrient uptake after ploughing 1288E
Response to minor nutrients 219E

SOWING AND TRANSPLANTING

    Clipping and row spacing   1022E
    Depth of seeding   822E
    Spacing   262Fr, 274E, 1137E
    Sowing dates   1352E

WEED CONTROL

    Dimethyl arsenic acid   1295E
    Pre-emergence weedicides   1040E

DISEASES

    Acrothecium Penniseti n. sp   824E
    Curvularia leaf blight   966E
    Curvularia lunata: India   14E
    ---: effect on grain   778E
    Dactuliophora   669E
    Ergot
        alkaloid synthesis   144Ge
        alkaloids of Pennisetum   18E
        control by fungicides   1259E
        meteorological features associated with   1066E
    Eye Spot   914J
    Foot rot disease in wheat   1253E
    Helminthosporium: India   223E
    ---: setariae   1382, 1384E
    ---: turciceim H. maydis
    ---: and H. carbonum   1106E
    Hypsoperine acronea   255E
    Microxyphiella Hibiscifolia: India   1203E
    Puccinia penniseti   1059E
        chemical control   1207E
        microbial control   1208E
        physiologic specialization   284E
    Rice and corn leaf gall virus   13E
    Root rot of cotton   831E
    Rusts   1049E
        control by chemicals   1207E
    Sclerospora graminicola   588E, 645E, 1129E, 1201E, 1130E,
        1270E, 1307E, 1308E, 1309E, 1330E, 1392E, 1436Ch
    Seed-borne fungi   1385Ge
    Sphacelia   25E
    Stringa lutea control with hormones   779E
    Sugary and green ear disease   1178E
    Tolysporium penicillariae   297, 494E
        control by chemicals   1383E
        Senegalense   477E
    Top rot   1054E
    Uredinearum   1275E
    Xanthomonas Rubricineans   1266E
    X. rubii sorghi   1076E

PESTS

Ceylon 495E
India 517, 573E
Tanzania 457E
Acrididae
    Senegal and Niger 757Fr
    Sudan and Chad 994E
Anthrodes sp. 1073E
Armyworm resistance 677E
Busseola fusca 674Fr
Caterpillars mining cereals in Tropical Africa 48Fr
Colemania sphenaroides 1252E
Contarinia sorghicola--East Africa 385E
Dysdercus superstituosus 384E
Empoasca Walsh--India 1024E
Heterococcus nigeriensis--Northern Nigeria 456E
Hieroglyphus nigrorepletus--India 1116
Locusta migratoria migratoriodes--Niger 311Fr
Nematode parasites of Amsacta moorei 126E
Schistocerea gregaria
    Eritrea 536It
    India 122E
    Sudan 586E
Sesamia cretica (S. pecki) Sudan 1290E
Stemboring agrotidae 142E
Storage pests of cereals
    East and Central Africa 944E
    Nigeria 3, 398, 399E
Tanymecus abyssinicus--Eritrea 535It
White grubs--biology and control--India 308E

FODDER 907, 1378E

Comparison of: Gahi and Starr 51E
---: varieties for dairy cattle 97, 251, 975E
Dwarf millet 16, 552E
Epistasis 197E
Fertiliser response for hay 92E
Gahi-1 459, 487E
Lattice design for forage yield trials 185E
Management: 189, 191, 328, 1021E
    Saudi Arabia 1229E
    Australia 143, 606E
    U.S.A. 823E
Napier-bajra hybrids 557E
Production: effect of cutting height and frequency 99E
---: under different moisture levels 115E
Quality
    affected by maturity 187, 196E
Silage trials 153E
Supplementary grazing 222, 1261E

FODDER (cont.)

  Chemical Composition and Nutritive Value   190E
    Digestibility   458E
    Feeding value   810E
    Harvest treatment--effect on forage nitrates   715E
    Pasture with buffers   809E
    Pusa giant Napier grass   550, 663E
    Varietal differences   240E
    Yield and composition affected by harvesting methods   163E

GRAIN

  Chemical Composition and Nutritive Values   552E, 1156Fr,
      1404E
    Calcium and phosphorus availability   403E
    Caretenoid content   282E
    Cariogenic component of experimental rat diets   714E
    Hybrids   1050E
    Indian diets   455E
    Oil   11E
    Straw   956E
    Sugar and nitrogen content   876F
    Threonine and arginine content   86E

UTILIZATION

    Algeria   383Fr
    As a mulch in tropics   839Fr
    India   272, 726, 814, 819E
    Niger   871Fr
    Sudan   1312E
    U.S.S.R.   578Rs

                        SETARIA ITALICA

MORPHOLOGY AND HISTOLOGY   207Sp
    Development of bristles   897E
    Development of caryopsis   782J

PHYSIOLOGY AND BIOCHEMISTRY
    Biology of flowering   485J
    Development phases and growth processes   565Rs
    Effect of protein on nitrogen retention   375E
    Effect of proteins on hemoglobin formation   377E
    Heat stress effects   326E
    Increasing salt tolerance   948Ch
    Liberation of amino acids from proteins   376E
    Photoperiodic response   829Bu
    Seed germination   518J
    Water requirements for germination   547Ch

GENETICS, CYTOGENETICS AND CYTOLOGY
    Genetical studies 1282J
    Interspecific crosses 680Ch, 681Ch
    Non-lodging mutant 1091E

BREEDING AND VARIETIES 834Rs, 1247E, 1280J
    Behaviour of populations and selections 88Sp
    Discriminant function in selection 1092E
    Hybrids 1263Rs
    Improved variety Navane 1233
    Isolation of varieties 27It
    Rust resistant varieties 574E
    Seed testing 1410E
    Variability in Kangni-1 71E
    ---: Kangni-2 1200E
    Variety trials 278Hu

AGRONOMY
    Effect of soil colloids on boric acid activity 397E
    Fertiliser experiments 977E
    Vegetable mulches 422E
    Weed control 236Rs

    Geographic
       Kenya 1148E
       Philippines: performance tests 1223E
       ---: yield trials 549E
       U.S.A. 1362E
       U.S.S.R. 50, 972, 985, 1240Rs

DISEASES 644Gr
    Bacterial leaf spot--India 1077E
    Biological aspects of smut 1364Rs
    Chaetochloa lutescens 117E
    Downy mildew 1286
    Grain smut 253E
    Helminthosporium graminicolous sp. in Japan 521J
    Helminthosporium species, H. maydis and H. carbonum
       1106E
    Infecting with S. destruens and u. crameri 643Ge
    Phytopharmacology of rice diseases 464Ge
    Phytophthora macrosporo 1291J
    Piricularia setariae 406E, 648E, 684Ch
    Rice blast fungus 916, 918J
    Seed borne fungi 425E
    Uromyces setaria-italica 893, 1051E
    Ustilago crameri 1267E

PESTS
    Azazia rubricans--India 1029E
    Colemania sphenaroides--India 1252E
    Hieroglyphus nigrorepletus--India 1116E
    Injuries by insect pests 655E
    Leptocorisa varicornis E 1169E

183

1127, 1203, 1254, 1266,
1269, 1270, 1271, 1301,
1302, 1303, 1348, 1349,
1355, 1360E
Fertilizers and plant nutrition
121, 414, 575, 771, 788,
817, 869, 1078, 1087E
Fodder 336, 770, 907, 954,
955, 956, 1044E
Genetics 637, 639, 1093E
Insects and control 122, 308,
361, 517, 573, 1024, 1029,
1116, 1252, 1255, 1333E
Medicinal plants 603E
Millet improvement programme
508, 509, 510, 511, 512E
Nutritive value 86, 94E
Taxonomy 272, 437, 725, 814,
899, 1379, 1421E

Israel
Diseases 588E

Italy
Breeding 1095It
Bronze Age millet 1310It
Diseases 427It

Japan
Diseases 521, 914J
Echinochloa 827J
Nutrition 1132J

Kenya
Agronomy 108, 589, 1147E

Madagascar
Breeding 209E

Mauritius
Fodder production 47E

Malawi (Nyasaland)
Agronomy 304, 929, 930,
1098E
Diseases 668, 1397,
1398E

Malaysia (Malaya)
Cultivation 1124, 1232E
Utilization 162, 165E

Mexico
Agronomy 214E

Morocco
Breeding 974Fr
Utilization 911Fr

New Guinea
Breeding 152Fr

New Zealand
Breeding 1276E
Diseases 281E

Niger
Agronomy 333Fr, 769E
Breeding 218, 870, 871Fr
Pests 311, 757Fr

Nigeria
Agronomy 799, 909, 1377E
Diseases 239Fr
Fodder 945E
Pests 3, 400, 401, 456E
Research 2Fr
Utilization 365E

North America
Agronomy 58E
Diseases 360, 1237E
Ecology 928E

Orange River Colony
Agronomy 718E

Pakistan
Coix lachryma-jobi 139E

Peru
Agriculture 768Sp

Philippines
Agronomy 549, 982, 1223E
Utilization 982, 1388, 1389,
1390E

Poland
Agriculture 632Pl
Plant remains 938Pl, 1399Pl

Puerto Rico
Agronomy 1364E
Physiology 90E

INDEX TO REPORTS
OF AGRICULTURAL DEPARTMENTS

C. S. I. R. O.
   (1961) 258
   (1963-1964) 101
E. Peru and Tingo Maria Agric. Exp. Station
   (1951) 768
Gorjkii State Agric. Res. Sta. (1953) 210
Manitoba Dept. Agric. (1953) 764
Matopos Res. Sta. (1956) 585
Overseas Food Corp. Tanzania (1955) 943
Rept. Admin. Ceylon (1960) 220
Rept. Agric. Dept. Mysore
   (1950) 865
   (1956-1957) 867
Rept. Agric. Survey Allahabad (1929) 562
Rept. Agron and Seed Div. Orange River Colony
   (1908-1909) 718
Rept. Agron. Dept. of Agric. Allahabad (1955) 391
Rept. Dept. Agric. Bombay
   (1922-1923) 166
   (1923-1924) 167
   (1936-1937) 231
   (1937) 407
   (1937-1938) 569
Rept. Dept. Agric. Madras (1941) 724
Rept. Dept. Agric. New South Wales (1967) 904
Rept. Dept. Agric. Northern Rhodesia (1949) 1098
Rept. Dept. Agric. Tanzania (1926) 1105
Rept. Dept. Agric. Uttar Pradash (1947-1948) 313
Rept. Manfredi Agric. Exp. Sta. (1954) 763
Rept. Massachusetts Hatch Sta. (1891-1892) 772

INDEX OF SERIALS
An index to all the entries herein that cite
a given periodical

Acta Agraria et Silvestria. Krakow. Series Agraria 1153
Acta Agraria et Silvestria. Krakow. Series Rolnicza 1152
Acta Agriculturae Sinica 225, 492, 1419
Acta Botanica Sinica 664, 948, 1411
Acta Pedologica Sinica 683
Acta Phytopathologica Sinica 684
Acta Phytophylacica Sinica 489, 1387, 1436
Advancing Frontiers of Plant Sciences 767, 1202
Agra University Journal of Research, Science 230
Agricultura Tropical 276, 370
Agricultural Bulletin. Federated Malay States 1232
Agricultural Gazette of New South Wales 15, 143, 469, 590, 1336, 1367
Agricultural Journal of India 1224, 1331, 1422
Agricultural Meteorology 104
Agricultural Research (India) 219, 500
Agricultural Research (Pretoria) 150, 1096
Agricultural Research (Washington) 16, 17, 181, 1328
Agriculture and Animal Husbandry in India 1125
Agriculture and Livestock in India 93, 1126
Agriculture (London) 530
Agrobiologiya 27, 408, 776, 1020, 1118, 1172, 1420, 1431
Agrokhimiya 940

L'Agronomie Tropicale 6, 46, 48, 119, 218, 227, 299, 333, 352, 757, 769, 998, 999, 1068, 1256, 1358
Agronomy Abstracts 1016
Agronomy Journal 98, 99, 163, 170, 190, 251, 356, 422, 449, 459, 715, 973, 1218, 1278
Allahabad Farmer 391
American Chemical Journal 660
American Journal of Botany 58, 169, 228, 695, 1380
Anais da Faculdade de Medicina da Universidade de Sao Paulo 305
Anais do Congresso Nacional. Sociedade Botanica do Brasil 95
Analyst 369
The Andhra Agricultural Journal 23, 874, 875, 876, 883, 1046, 1076, 1081, 1082, 1084, 1134, 1135, 1354
Angewandte Botanik 608
Annales de la Nutrition et de l'Alimentation 211
Annales du Centre de Recherches Agronomiques de Bambey Au Senegal 135, 528
Annales Universitatis Mariae Curie-Sklodowska. Sect. E. Agriculture 756
Annali della Sperimentazione Agraria 421
Annals of Applied Biology 647
Annals of Arid Zone 1, 302, 766, 818, 822
Annals of Botany 396
Annals of Missouri Botanical

Index of Serials 193

COLLOQUIAL NAMES FOR MILLET SPECIES*

1. Brachiaria ramosa Stapf [Panicum ramosum Linn.]**
   browntop millet

2. Coix lachryma-jobi Linn.
   Job's tears, adlay.   Dutch: djalibras

3. Digitaria iburua Stapf; and, D. exilis Stapf
   fonio, hunry rice

4. Echinochloa colona (L.) Link.
   jungle rice, shama millet
   Echinochloa frumentacea (Roxb.) Link. [Panicum frumentacum
   Roxb.]
   Japanese barnyard millet, billion dollar grass.   Indian:
   Kudiraivali, Sawan.   Chinese: Kweichou

5. Eleusine coracana Gaertn.
   finger millet, ragi.   Indian: ragi, kangni, nachani.   Italian: rapoke

6. Eragrostis tef (Zucc.) Trotter [E. abyssinica]
   teff

7. Panicum miliaceum Linn.
   proso; and common, broom, bread, or hog millet.   Russian: prosa.   German: Kolbenhirse

8. Panicum miliare Lam.
   little millet

9. Paspalum scrobiculatum Linn.
   Indian: kodra, varagu, kodon

10. Pennisetum typhoides Stapf and Hubb.
    Bulrush or pearl millet.   Indian: Bajra, bajri.   Tamil:
    Cumbu.   Arabic: dukhun.   French: mil a chandelles

---

*English language names given first; other languages are identified.
**Synonyms of scientific names used for the same species are
bracketed.   However, there are two distinct species of Digitaria and
Echinochloa.

200

11. Setaria italica Beauv.
         Foxtail; and Italian, German, or Hungarian millet.   Indian:
     tenai, navani.   Russian: mohar

12. Sorghum bicolor (L.) Moench [S. vulgare Pers.]
         sorghum, durra, Kaffir corn, guinea corn.   Indian: jowar
     jola, Tamil: Cholam.   Arabic: dura

LANGUAGE ABBREVIATIONS

These language symbols follow the item numbers in the Author, Species, and Geographical Indexes and show the language of the original text of the reference.

Af	Afrikaans
Ar	Arabic
Bu	Bulgarian
Ch	Chinese
Cz	Czech
Du	Dutch
E	English
Fr	French
Ge	German
Go	Georgian
Hu	Hungarian
It	Italian
J	Japanese
Pl	Polish
Po	Portuguese
Rm	Rumanian
Rs	Russian
Sc	Serbo-Croatian
Sp	Spanish
Uk	Ukranian
Yu	Yugoslavian